Plastics and the Environment

Editors:

Frances Gardiner

Eleanor Garmson

SMITHERS *i*Smithers

*i*Smithers – A Smithers Group Company

Shawbury, Shrewsbury, Shropshire, SY4 4NR, United Kingdom
Telephone: +44 (0)1939 250383 Fax: +44 (0)1939 251118
http://www.ismithers.net

First Published in 2010 by

***i*Smithers**

Shawbury, Shrewsbury, Shropshire, SY4 4NR, UK

A catalogue record for this book is available from the British Library.

ISBN (Hard-backed): 978-1-84735-491-4
ISBN (Soft-backed): 978-1-84735-556-0
e-book: 978-1-84735-557-7

Typeset by *Argil Services*
Indexed by Argil Services
Printed and bound by Lightning Source

Preface

For many in the plastics sector, environmental sustainability may be part of a wider corporate social responsibility scheme with the primary purpose of increasing customer loyalty and penetrating new markets. However, everyone knows that there are also real opportunities for companies to gain substantial cost savings - not to mention environmental integrity - through efficient processing and resource management.

In the past so called 'green' materials were simply not able to meet the aesthetic demands of customers or to achieve the required level of durability and functionality in plastic products, and very often recycled materials were more expensive to produce than their virgin equivalents. Not so now. New application ideas, novel designs and lower cost material solutions are now readily available and some are discussed in this book.

This Update aims to address the environmental decisions that the non-packaging plastics supply chain can take to increase their environmental integrity, and to comply with current and pending regulations.

With the plastic products sector acknowledging the need to consider the environment throughout a product's life cycle, many companies now recognise the benefits of communicating these efforts to the consumer through product labelling to encourage a purchase and simplify post-use disposal.

This book features best practice cases from key industries to illustrate the steps that can be taken at each stage of product development – including conception, and design and materials selection.

This book will be of interest to all those who produce both plastic raw materials and to those who produce plastic products.

The Editors would like to thank all the Authors for their patience and co-operation during the production of this book.

Frances Gardiner
iSmithers
August 2010

Eleanor Garmson
iSmithers
August 2010

Plastics and the Environment

Contents

1 Developments in Polymer Technology Driven by the Need for Sustainability

John A. Colbert

1.1 Introduction

Are our developments in sustainability driven by our conscience or by the high cost of energy? Perhaps it is both. Can we solve everything just by banning the use of plastic bags, or should we actually use more plastic? It is not easy to satisfy both public perception and the scientific community. Different European Union (EU) funded research projects are examined, all of them partly, or wholly driven, by the need for energy reduction within the industry. Together they cover the understanding of energy, benchmarking, legislation, new materials, and new processes.

1.2 What Drives Developments Forward?

Firstly we need to understand what it is that drives developments forward. It is often said that developments progress faster in times of conflict, or when a crisis occurs. In some ways we could describe our current environmental situation as a time of crisis. We are a high carbon economy, burning, consuming, destroying, and polluting, see **Figure 1.1**. But we are also devising ways and means to cut carbon emissions, use resources more efficiently, and increase our recycling capability. One could say that we are all at war with ourselves - the consumer within us against the environmental conscience. So, is it our awareness of the need for sustainability that drives our developments, or is it money? Are our developments really driven by our conscience or is it the high cost of energy? In reality it is both, and we need to be pragmatic and recognise that most companies will aim to reduce their energy consumption in order to reduce costs, maximise profits, and stay alive. The fact that they can then take the moral high ground as well is a bonus (**Figure 1.2**).

Figure 1.1 Burning, consuming, destroying, and polluting

Figure 1.2 The moral highground

1.3 How can we save the World?

Having destroyed the myth that we are all trying to save the planet, we now need to decide what we can do to reduce the environmental impact of our industry, save the world, and increase our profits. Do we follow the public perception, fuelled by the mass media, and cut out our use of the humble plastic bag? Some people believe that this will solve everything! What is the reality? In the UK we use around 230 bags per person per year which contributes the same carbon emissions as driving 25 km in our car! Or should we use more plastics? Colin Williamson of Smile Plastics maintains that 'plastics are the environmentalists friend', and one of the facts that he cites is that the Airbus 380 uses 25% plastics and it consumes 30% less fuel than a Boeing 747.

1.4 Getting the Science right

At Smithers Rapra we carry out Research and Development on a wide range of polymer related issues, often these are European projects funded by the EU, and carried out in co-operation with various partners throughout the European community. Two of these projects, funded by 'Intelligent Energy', have involved a wide range of partners across the EU, and deal very much with sustainability. The first one: Reduced energy consumption in polymer engineering – **RECIPE** (*www.eurecipe.com*), looked at the energy usage for the major plastic converting processes, and published a guide to '*Low Energy Plastics Processing*', of which around 3,000 copies have now been downloaded from the website (**Figure 1.3**).

Figure 1.3 Low energy best practice guide

RECIPE, which finished a three year programme in 2007, also produced a downloadable programme aimed at calculating the cost of ownership for injection moulding, and an online programme that enabled companies to benchmark their factory's energy consumption, and produce action plans specific to their operation for reducing energy consumption. The basic premise is that you cannot control something unless you can measure it. Therefore, in order to reduce energy consumption, we need to measure first what we are currently using and then measure any changes, which hopefully will be reductions. We also need to be able to set ourselves targets, which is where benchmarking becomes important. **Figure 1.4** shows the comparative energy usage for a variety of plastic conversion processes, including the energy involved in the associated infrastructure.

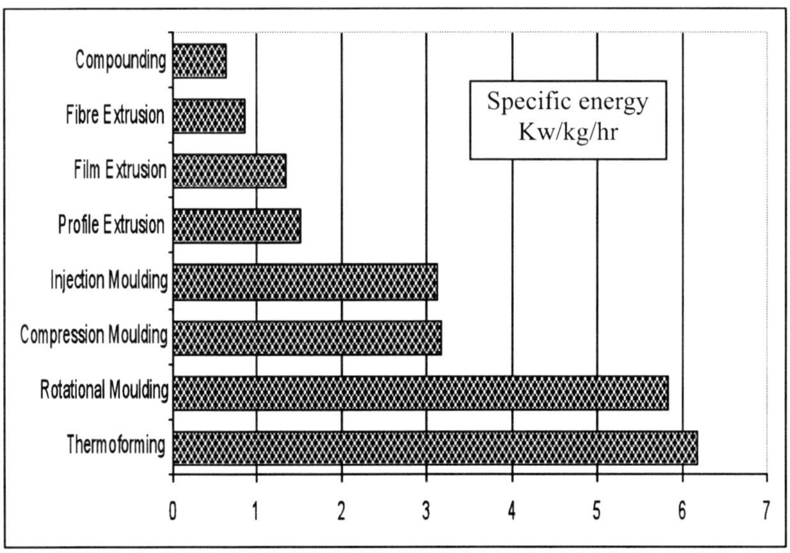

Figure 1.4 Specific energy of key processes

1.5 Legislation and Design

Following on from **RECIPE** is another project entitled **ENER-Plast** (*www.enerplast.eu*) which offers a wide range of information to assist companies in moving towards an energy efficient operation. The first output was a guide to environmental legislation throughout the EU, covering all the member states and listing the different pieces of legislation enacted in each state, or at least all that could be found. One of the facts highlighted was the wide disparity between different states in terms of the number of laws and their complexity. One could say that a lot of energy needs to be expended in order to comply with all the energy legislation. There will also be available on the website a carbon calculator to enable organisations to estimate their own carbon footprint. Finally there will be a guide to environmental design, and a list of energy efficient machinery. The aim of **ENER-Plast** is to offer an overview of the carbon footprint of the whole polymer supply chain.

The key principles, from these energy driven projects, are currently being developed into a training programme that will be available across the European community. This is within a 'Leonardo' funded project entitled **'ENERGYWISE Plastics'**. The focus of the project will be the development of an e-learning platform and supporting handbook.

These three projects, **RECIPE**, **ENER-Plast** and **ENERGYWISE**, cover the whole polymer supply chain (**Figure 1.5**) and effectively look at what we currently do, how we can do it in a more environmentally friendly way, and training to ensure that everyone becomes aware of how to save energy.

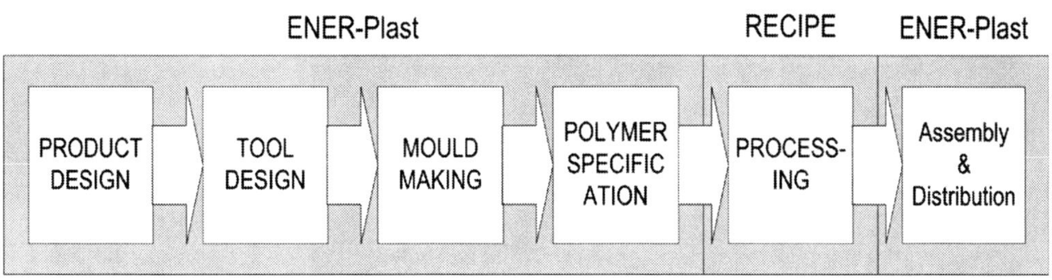

Figure 1.5 Area covered by 'RECIPE' and 'ENER-Plast'

1.6 New Materials

Now let us consider the development of new materials. There are two main projects in this area: **PolyCond** (*www.polycond.eu*) and **Pegasus** (*www.pegasus-eu.net*), both European projects funded through the EU Framework 6. Firstly **'PolyCond'**, which concluded in 2009, where the objective was to develop a conducting polymer. We know that by adding large amounts of carbon black (30% or more), to a polymer we can obtain a degree of conductivity, but at the expense of a reduction in physical properties. This project developed a technique for incorporating small amounts (less than 5%), of carbon nanotubes into the polymer matrix and then a range of grades was developed, based upon different materials, and a range of case studies showing real applications. The ability to use an injection mouldable polymer having inherent shielding and electromagnetic interference properties meant that there was no need to use metallising and hence there was a large reduction in the energy required to manufacture the component. Most of the case studies were able to show reduced energy consumption, which was essentially one of the drivers in the development of this new material.

The 'earthing plug', shown in **Figure 1.6**, utilises a conductive polymer for the earth pin, the end result being 80% weight reduction, 40% cost reduction, the ability to manufacture in small lot sizes, and significant energy reductions.

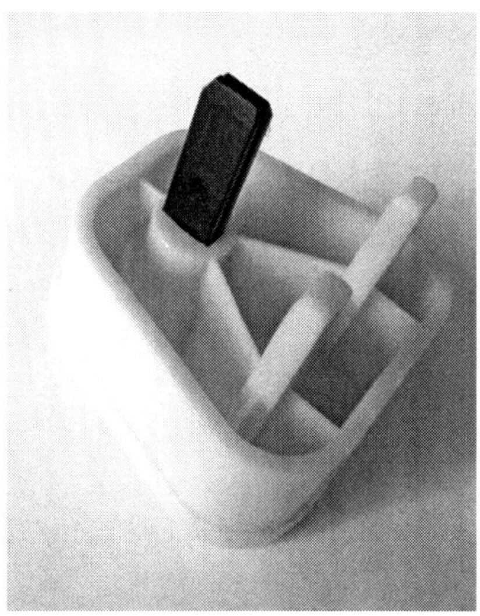

Figure 1.6 Non-metallic earthing plug

'**Pegasus**' is an ongoing project on two different levels. The first is the adoption of a new philosophy and software in the motor industry to speed up development of new components and the second is a range of polymer developments aimed at the motor industry. In the motor industry the use of plastics in the average car saves 750 litres of fuel over the car's lifetime, making it roughly 5% to 10% more efficient.

With the new materials developed within the '**Pegasus**' project we are aiming to improve on this energy efficiency. They include moulded in colour using nanocolour, and materials that will de-bond on command, thus making dis-assembly and recycling of the car more efficient. A demonstrator, incorporating many of these features in the rear wing of a new car, will be used at exhibitions and shows to highlight the innovations (**Figure 1.7**).

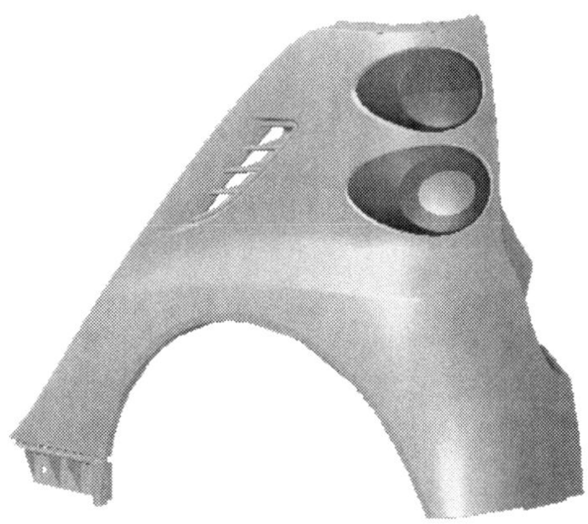

Figure 1.7 Rear wing demonstrator for 'Pegasus' project

Various moulding techniques are also being developed within '**Pegasus**', including back-moulding, over-moulding, local reinforcement and multi-functional moulds. All of these will generate environmental benefits, sometimes energy reduction but also re-use, re-cycling, and better use of resources.

1.7 New Processes

Finally, there is a whole range of projects that are aimed at process improvements.

Rotoflex (*www.rotoflex-eu.org*), an EU Framework 7 project, is developing energy efficient methods of producing multi-walled rotational mouldings, and **FlowFree**

(*www.euflowfree.com*) developed a process for injecting supercritical carbon dioxide into the melt stream to reduce the viscosity and hence the energy needed to process certain polymers. **Figure 1.8** illustrates the reduction in die pressure with polyvinyl chloride when different amounts of carbon dioxide were injected into the extruder process.

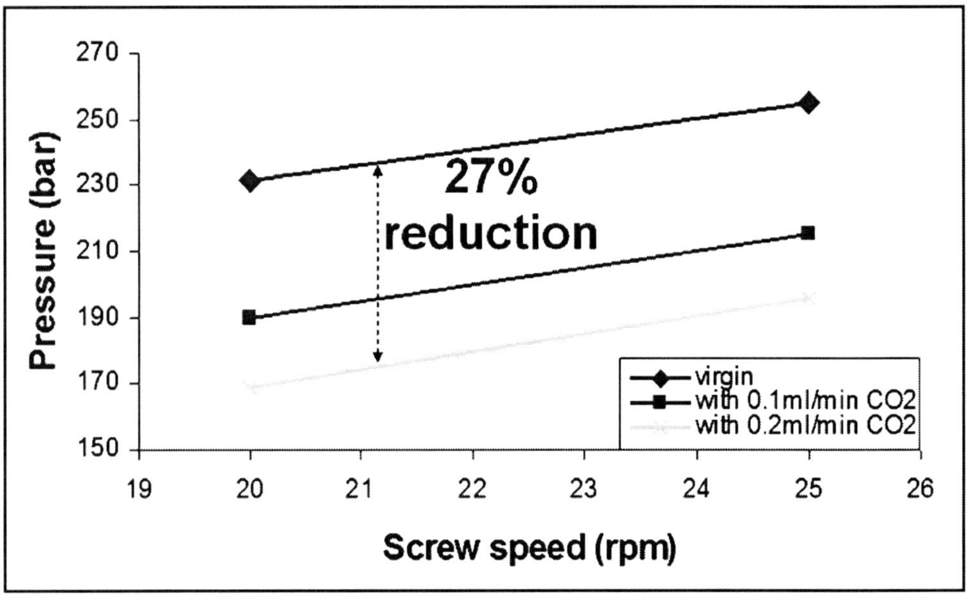

Figure 1.8 Pressure reduction when using CO_2

There are also two projects: **DevulCO2** (*www.devulco2.net*) and, following on, **ReMould** (*www.remould.org.uk*) that are successfully devulcanising waste tyre rubber back to a high quality mouldable state and are now developing the process to a commercial level: true recycling as opposed to reuse at a lower level. These are both funded by the UK Technology Strategy Board.

Finally, a technique now being used within **PeptFlow** (*www.peptflow.com*) could well help to understand exactly what happens within polymer processing equipment and hence enable us to design more efficient machinery. This project means that, for the first time, the path of a particle of polymer flowing through different elements of a twin screw compounding extruder can be tracked in real time and whilst operating a real process. The technique is known as positron emission particle tracking and we are now in a position to quantify the mixing ability of different twin screw elements. This process could also be applied to various types of batch mixers as well as other types of extrusion equipment.

Figures 1.9 and **1.10** show the horizontal, axial and vertical position of the particle as it moves through the extruder window. By repeating this experiment many times statistical analysis of the flow through a wide variety of twin screw elements was obtained. The residence time distribution was evaluated along with other factors that enabled an assessment to be produced of both dispersive and distributive mixing effects.

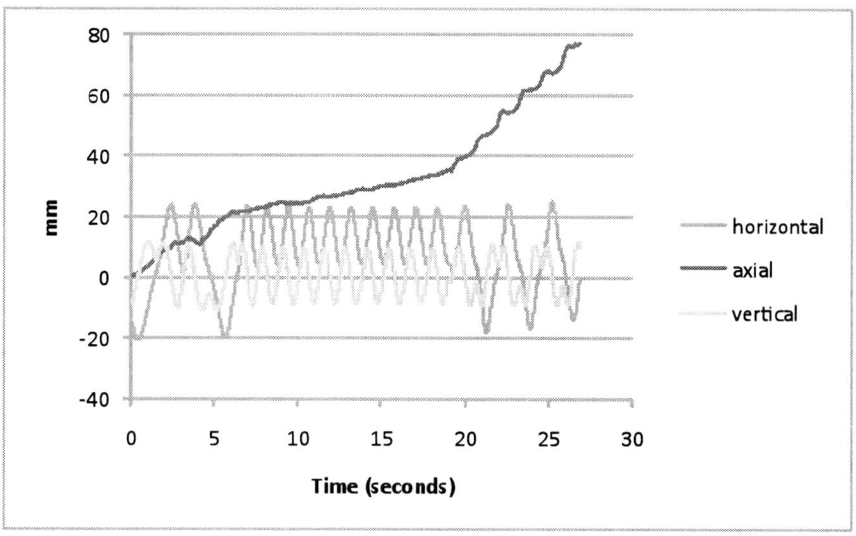

Figure 1.9 Trace of PeptFlow particle

A design booklet has been produced as part of this project which summarises the findings and offers a guide as to which elements gave the best distributive or dispersive mixing effect.

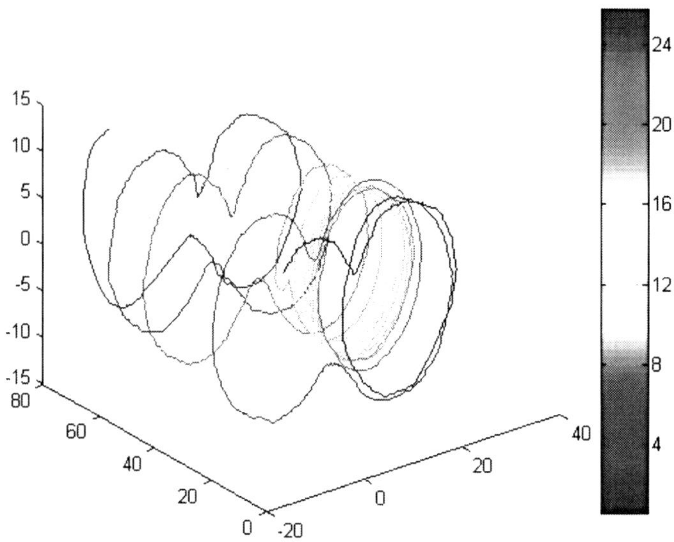

Figure 1.10 3D view of PeptFlow particle

1.8 Conclusions

In conclusion therefore, we need to reduce our carbon emissions to save the world for our children and for their children, but this does not mean using less plastic. In fact it almost certainly means using more plastic to make us more efficient in our energy usage (using 1 kg of oil to manufacture insulating foam for a domestic dwelling will save 3 kg of heating oil). Either our desire for sustainability, or our need for improved profitability, are driving developments forward, but they are both currently driving in broadly the same direction, so let's make the most of it! EU funded projects are not only bringing European research together, but are acting as a catalyst for major developments in polymer technology, both developing new materials and improving processing techniques.

Bibliography

1. C. Williamson in *The World of Plastics*, Guardian Newspaper Supplement, Smile Plastics, 2009.

2. A. Azapagic, A. Emsley and I. Hamerton in *Polymers, The Environment and Sustainable Development*, John Wiley & Sons Ltd., Chichester, UK, 2003.

3. V. Goodship in *Introduction to Plastics Recycling*, Rapra Technology Ltd., Shrewsbury, UK, 2007.

4. *The Sustainability of Plastics Products*, European Plastics Converters, Brussels, Belgium.
http://www.plasticsconverters.eu/uploads/webpresentation%20 Communication.pdf

2 A Medium Voltage Switchgear Mechanism which is Insensitive to its Environment

Romain Maladen, Christophe Prévé, Daniel Piccoz, Patrick Bailly and Bernard Valette

2.1 Introduction

The Transmission and Distribution (T&D) division of AREVA offers a full range of products and services intended to transmit and distribute the electricity from power plants to final users. Among these products, the medium voltage switchgears that we generally used in the secondary distribution networks are designed and manufactured in the factories located in Mâcon (France), and in Guardamiglio (Italy). AREVA T&D, Mâcon currently offers two different kinds of medium voltage switchgears with switch-disconnectors to its customers: the Fluokit M24+ and the FBX (**Figure 2.1**).

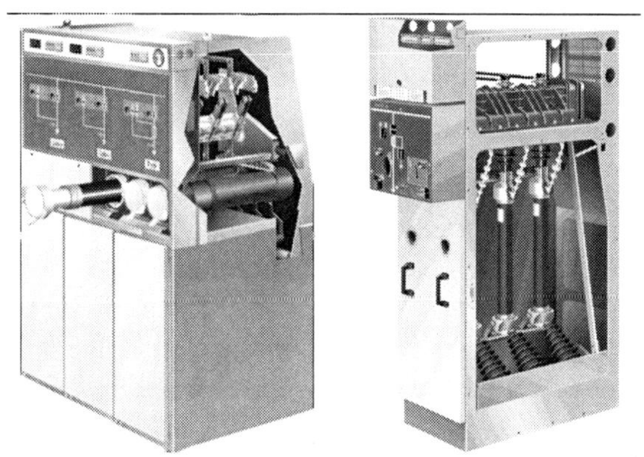

Figure 2.1 Medium voltage distribution units manufactured in Mâcon

Despite their difference these two units have a common point: their switches, which require high speeds and high forces, are driven by switchgear mechanisms. These mechanisms are therefore essential and for that reason have to be reliable and efficient over many years (up to 50 years according to international standards). AREVA, like its competitors, has always been building them mainly with metals (**Figure 2.2**). However, despite their interesting mechanical properties, the metallic parts present some drawbacks such as corrosion or grease alteration (downgraded) for example. Solutions intended to suppress these drawbacks exist but are not always reliable and are not environmentally friendly: the surface treatment employed against corrosion contains noxious products and the lubrication of the mechanisms to facilitate their operation needs regular maintenance operations.

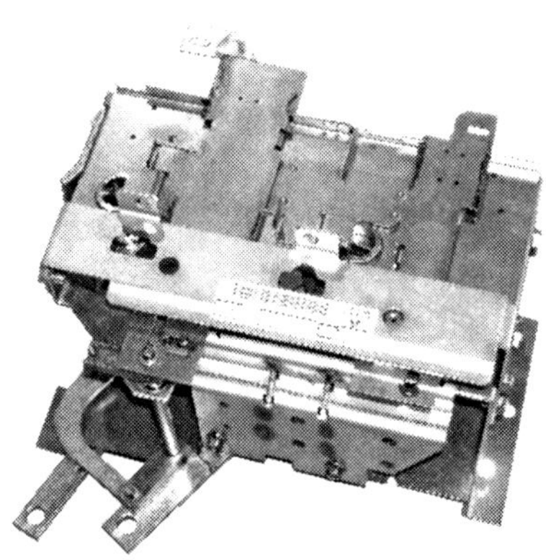

Figure 2.2 Example of current switchgear mechanism

Our continuous improvement process always leads us to find a way to improve our products in terms of reliability, lack of maintenance, and environmental respect. Unfortunately, for the switchgear mechanisms, it was realised that it would be impossible to reach these different targets with mechanisms made from steel. It was concluded that our only solution was to develop new technology and to investigate a switchgear made from polymers. Indeed, the polymer materials are not damaged by the environment and are recyclable, they don't need to be lubricated and they allow complex parts to be designed with a reduced environmental impact compared to other technologies. Thus, it was decided to investigate this technology and to verify its environmental benefits with a life cycle assessment (LCA).

2.2 Selection of the Most Appropriate Material

Engineering thermoplastic materials are relatively young and their ageing is not always well known and/or not always sufficient. Consequently, the first challenge was to check which materials could keep their mechanical properties for more than 50 years. In order to identify the best material for our application and then to predict its life duration a study was started in collaboration with a reputed laboratory specialised in polymers' ageing: the 'Laboratoire d'Ingéniérie des Matériaux' of the ENSAM, Paris, France. Their methodology to predict life span couples the analysis of structural changes at all pertinent levels with advanced polymer physics. In collaboration with the laboratory, AREVA T&D chose a high-strength, semi-crystalline engineering thermoplastic often marketed as a metal substitute for corrosive environment: polyoxymethylene. After a three year study, the conclusions given by the laboratory allowed us to envisage the design of a thermoplastic made mechanism.

2.3 Design of a New Range of Mechanisms

First of all a characteristic part of a current mechanism was designed to validate the technology. The part called a 'tumbler' was chosen because it has complex shapes and it is the most constrained part of the mechanism: static forces and shocks.

The challenge was important because the design of thermoplastic parts with high mechanical constraints was unusual for us and we had to take into account all the constraints linked to this technology.

After the first draft, the new model was improved using numerical simulations and then, when the shapes were validated, a prototype was launched and successfully tested.

Gradually, all parts were designed and finally a completely new mechanism was born (**Figure 2.3**). The first tests with mechanisms from the try-out confirmed excellent results as they completed several thousands of operations while international standards request only 1,000 operations.

At the end of this first study and thanks to its excellent results, other mechanisms were designed. In the future a full 'Compodrive' range will be available.

Figure 2.3 Example of new mechanism

2.4 Environmental Studies

In harmony with AREVA T&D, the industrial plant of Mâcon has an environmental protection policy. It is ISO 14001 [1] certified, also including the design of products since 2002. An environmental management system is in place on the site and procedures concerning the design of new products have been introduced within the context of the eco-design. From this, AREVA T&D have acquired much experience in this field for many years and a network of more than 100 experts of eco-design is currently in place in the group. Each subsidiary of the AREVA product line is evaluated every year and a wide range of people such as designers, project managers, sourcing engineers and sales managers are aware of eco-design. A document called *the Green Guide* was also written. It is a tool intended to help the designers to check if all the points regarding the environmental impact of a future product were taken into account during the design phase. This document also includes design rules such as forbidden materials or surface treatments, the best way to assemble several parts regarding their future dismantling and so on.

In the following sections a study is presented which compares the environmental impact of two kinds of medium voltage switchgear mechanisms by analysing their material balance and their LCA realised with a specific software called 'SimaPro'. These two products were studied with the same tools, same methodology and same hypothesis so that the comparison was valid. The first product studied is the current steel mechanism and the second one is a new mechanism made with an engineering thermoplastic material. This second mechanism will soon replace the old one on all the products.

2.5 Material Balance Analysis

During the first step of an environmental study a material balance is prepared for the analysis phase of the LCA. In this case it was observed that the new thermoplastic made mechanism had almost half the number of parts compared to the old steel mechanism. Indeed this new product is made of 33 parts while the old one was composed of 58 parts. This first observation is important because it means that the new product will certainly be easier to dismantle, and subsequently to recycle, than the old one.

It was also observed that substituting polymers for steel allowed for a significant reduction in the mechanism's weight. For example, a CD10 mechanism (basic mechanism of the modular range) is three times lighter than its steel equivalent. It is currently made of 1.8 kg of steel and of 2.3 kg of thermoplastic materials while before it weighed 12.3 kg (mainly composed of metallic materials) (**Figure 2.4**). Similar figures were also obtained with the CD110 mechanism (basic mechanism of the compact range) as its weight was reduced from 8.9 kg to 4.6 kg. For the CD110 mechanism, the material balance between steel and polymer materials is similar to that for a CD10. In addition to the environmental aspects directly linked to the materials, that are discussed later, the weight reduction allows for the reduction of the impacts linked to the transportation and help to facilitate the mechanism assembly on the cubicle.

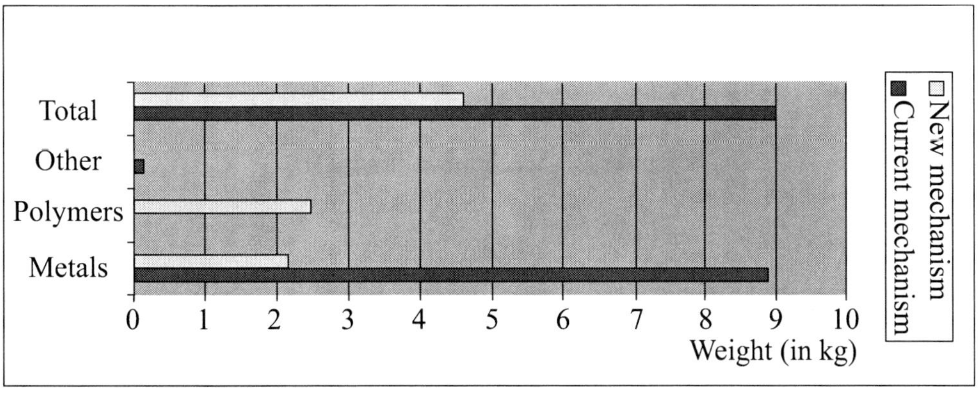

Figure 2.4 Material balance analysis

2.6 LCA

An LCA consists of studying the environmental impact of a product 'from the cradle to the grave' (**Figure 2.5**). It means that all phases are accounted for such as the mining of raw materials, the manufacturing of the product, its packaging, its transportation, its use, its end of life. The purpose of an LCA is to limit the product impacts on natural resources and on the environment (wastes, emissions of noxious products) by identifying the most important process. The final result of this kind of study is generally divided into several environmental 'impacts', for example, the global warming expressed in grams equivalent of CO_2.

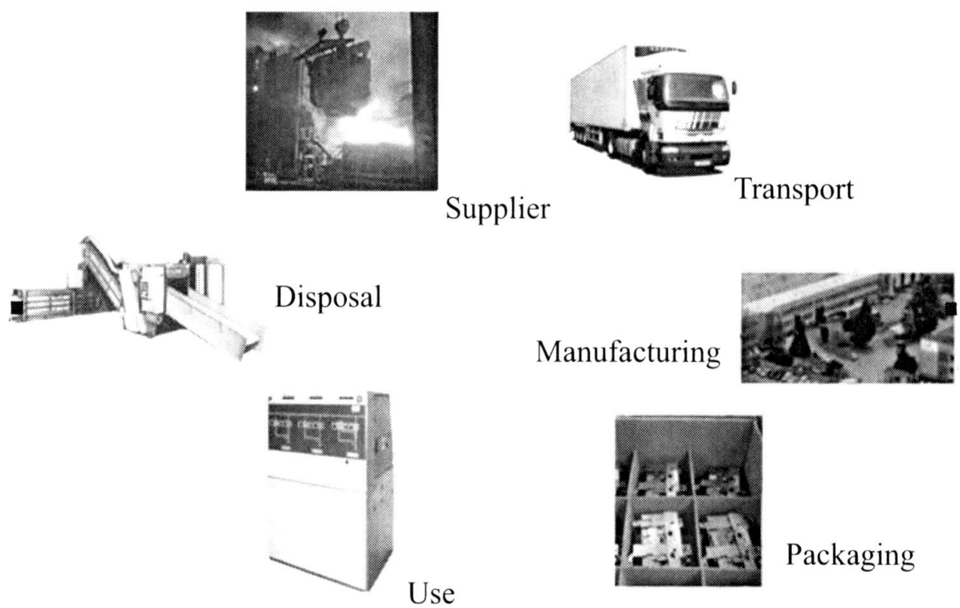

Figure 2.5 Mechanism life cycle

The first step of the LCA consisted in modelling the assembly of the mechanisms. This phase was done using to the collected data available in the detailed list of materials of each product. For each part a process associating its material, its manufacturing, its surface treatment and its eventual process of energy or transportation was defined. The parts were then organised in subassemblies and in assemblies. The whole life cycle is composed of these assemblies, an end-of-life scenario, several processes (transportation, energy) and complementary life cycle (packaging for example).

For the CD110 mechanisms, which were studied last, this study showed that the newly designed mechanisms were really better than the old ones from the point of view of environmental impact (**Figure 2.6**). Most of the environmental impacts were indeed reduced by half and some of them, such as photochemical smog, nuclear waste and air toxicity were reduced more significantly. Only global warming, acidification and resources stayed at a similar level. The only impact which is not improved of compodrive mechanisms is the production of hazardous wastes.

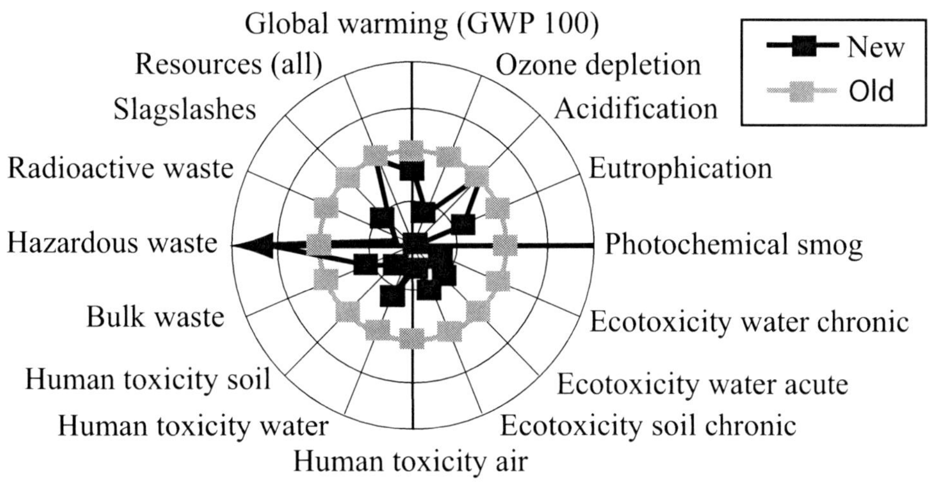

Figure 2.6 Comparison of environmental impacts

A more detailed analysis of these results showed that the manufacturing phase was the most important concerning its impact on the environment (**Figure 2.7**).

In order to communicate these results to people who are not always aware of technical jargon, an attempt was made to popularise these impacts. To do that, they were compared with products in everyday life. For example, the global warming impact of our mechanisms was compared with a classical 75 W electrical bulb. We showed that the new mechanism was equivalent to 100 full days of lighting of this bulb while the old mechanism was equivalent to 120 days. For ozone depletion it was shown that 62,000 new products were equivalent to one fridge while before the equivalent was 22,400 fridges.

This LCA was finally completed by additional studies: an end-of-life guide and an ecodeclaration leaflet were written and a technical and economic study about the recycling was also prepared.

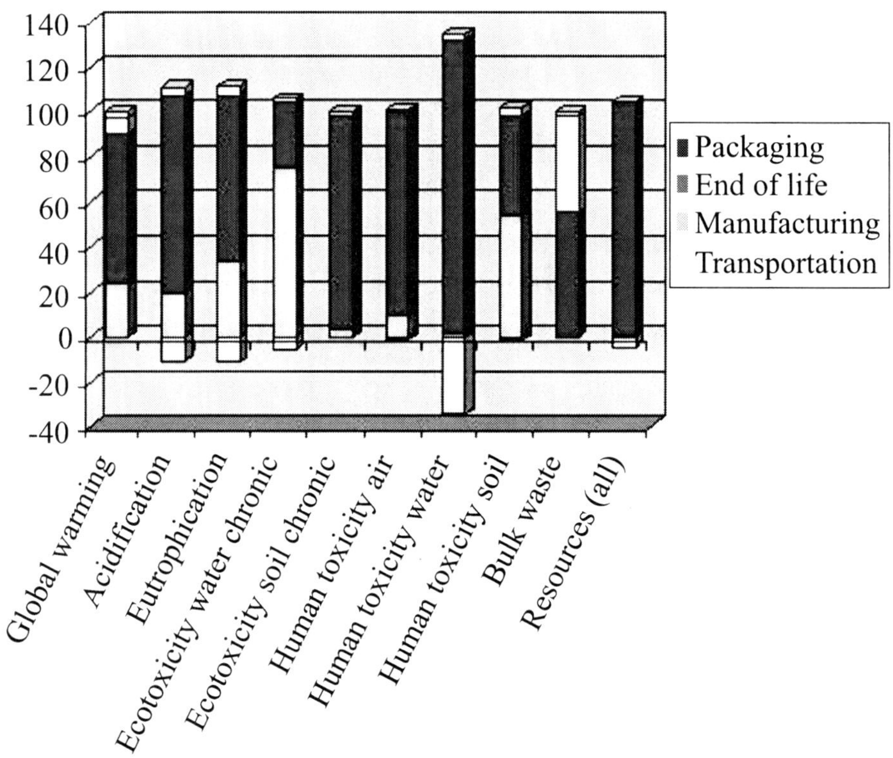

Figure 2.7 Process impacts

2.7 Conclusion

This new range of mechanism, prepared with a technology never used before for this kind of application, presents numerous advantages. 'Compodrive' mechanisms are more resistant to mechanical endurance and they are entirely insensitive to outside conditions with an added environmentally friendly touch. Indeed, maintenance operations and thus grease alteration (downgraded) are suppressed, their dismantling is easier and their materials have a lower environmental impact and are recyclable. The ultimate result will be further improved service continuity of electrical networks. In conclusion, the new 'Compodrive' range is more than an evolution, it is a revolution! Customers, like Electricité Réseau Distribution France, who have already discovered these new mechanisms, applauded the technological breakthrough.

References

1 ISO 14001, *Environmental Management Systems - General Guidelines on Principles, Systems and Support Techniques*, 2004.

Bibliography

1. B. Fayolle, J. Verdu, M. Bastard and D. Piccoz, *Journal of Applied Polymer Science*, 2008, **107**, 3, 1783.

2. B. Fayolle, J. Verdu, M. Bastard and D. Piccoz, *Journal of Applied Polymer Science*, 2009, **111**, 1, 469.

3. M. Bastard, *Etude de la Durabilité de Pièces Thermoplastiques, Application au Polyoxymethylène,* Mecanique et Materiaux, ENSAM, France, 2006. [PhD Thesis]

3 From Industrial Polymerisation Wastes to High Valued Material: Interfacial Agents for Polymer Blends and Composites based on Chemically Modified Atactic Polypropylenes

Jesús M. García-Martínez, Susana Areso, Jesús Taranco and Emilia P. Collar

3.1 Introduction

Functionalisation of polymers has been considered to be a good way to improve properties that would limit their commercial value. The latter, that is valid for plenty of commercially available and/or useful polymers, becomes more important when the polymer to be modified has a previous marginal value.

Isotactic polypropylene (iPP) is a very interesting polymer due to its high versatility that makes it to cover a wide range of applications (either as neat polypropylene (PP) or combined into hetero-phase materials). These applications, mainly due to its stereo-specificity, make it suitable for applications from the commodity polymer uses to the engineering materials. The consumption of this material has passed 10 million tons in the late 1980s to a value close to 44 million tons in 2005, and it is estimated that in 2010 the demand will increase beyond 50 million tons per year. The latter provides a clear idea of the importance of this material in the world global market. However, in the industrial production of iPP, a significant fraction of atacti PP (aPP) is also produced. This polymer exhibits a lack of stereo-specific regularity that confers on it a set of totally different properties in comparison to iPP. It can be said that the aPP is a tacky material, mainly amorphous that exhibits a low glass transition temperature (T_g = -22.4 °C) compared to iPP (close to 0 °C), and with serious lack of properties, mainly the mechanical ones to be considered useful as a material itself. In fact, aPP exhibits little strength and its main application is in conjunction with bitumen or asphalt [1]. However, it must be mentioned that for many years aPP was considered a waste rather than a by-product but nowadays it is used in jointly with asphalt in roofing applications, sealing strips, and acts as a provider of stability to asphalt surfaces, adhesives, and so on [2]. In fact, aPP is actually specifically produced for these applications by controlling the process for obtaining high molecular weight (M_w) polymer rather than the low molecular ones when it is a by-product. It is noteworthy to mention that the amount of it produced is strongly dependent on both the type of process used to obtain the stereo-specific polymers and of the catalyst system used [3]. Nevertheless, it can be said that, in dependence to the former, the generation of these wastes are in the range between 2% and 15% of the overall production of PP. Nevertheless, the applications for aPP (by-product) have been mainly restricted to marginal uses up to day. This work shows the option to convert this industrial by-product in the raw material to obtain interfacial agents. This will enhance the performance of polymer blends and composites by promoting the adhesion between the iPP and other polar polymers or mineral reinforcements (micro and nano reinforcements). The option to improve the properties of aPP by grafting of polar monomers to make it attractive for any uses was early suggested by Natta (discoverer of stereo-specific polymerisation of propylene) in late 1950s [4]. However, although the chemical modification of this polymer indeed enhances its properties, it is their use as interfacial modifiers in hetero-phase materials based on polymers which have been demonstrated to be highly effective. This option was not considered in the literature until the authors established that an interfacial agent with an amorphous character would be more easily allocated at the interphase between components in polymer composites, which resulted in better interfacial interactions [5-10]. To develop this concept authors modelled the behaviour of such heterogeneous systems in terms of Box-Wilson experimental designs, factorial, centered and rotatory [10]. Finally, the tremendous efficiency of some of the modified aPP obtained has proven to be better than even those of the leader interfacial agents in the market [11].

Following all the previous considerations this chapter shows, at a glance, the proven and real possibilities of modification of this waste. So, the chemical modification of aPP with grafting of a series of polar monomers such as maleic anhydride (MAH), succinyl fluoresceine (SF) and *p*-phenylene-bis-maleamic acid (*p*PBM) performed in solution and/or in the molten state has been reported [11-14]. The use of such different reaction media has allowed authors to define a chemical reaction path that explains the differences in architecture, structural properties and ultimate behaviour of the modified polymers [11-16] and the heterogeneous materials based on them, including micro- and nano-structured blends, alloys and composites which are in the vanguard of the so called advanced materials based on polymers [4-10].

3.2 Chemical Modification

The aPP waste used was supplied by Repsol-YPF (density = 0.85 g/cm^3; M_w = 54,000; number average molecular weight (M_n) = 2,700; T_g = -22.4 °C). To obtain the grafted aPP with succinic anhydride (SA) moieties and *p*PBM moieties (aPP-*p*PBM), a MAH and a *p*PBM acid (synthesised in our laboratories [14]) were used. As reaction initiator, a dicumyl peroxide (POx) was used. To prevent thermo-oxidative degradation, a series of thermal stabilisers were incorporated. Additionally, the aPP with grafted SF groups was obtained from aPP-SA by a condensation reaction between SA groups and resorcine in the presence of $ZnCl_2$ as a catalyst [5, 13]. The solvents used for both the process in solution and the washing operation of the modified polymers were xylene (isomer mixture) and dimethyl-formamide (additionally only for aPP-*p*PBM). **Figure 3.1** shows a scheme of these reactions. The chemical modification in solution was conducted at boiling solvent temperature in 1/30 and 1/5 *w/w* polymer/solvent ratio. Additional experiments were performed with MAH in a further reaction step with resorcine to yield aPP-SF. The reactions in the molten state were performed in the same way (and absence of solvent) in a mixing chamber (Rheomix 600 by Haake) working at different gear rates [11-16]. Experiments were performed according a Box-Wilson Experimental worksheet by considering POx in the 0.7 to 4.1% *w/w* range and the polar monomer in the 0.5 to 17.5% *w/w* interval. Operation temperature was fixed at 140 °C in all cases [11-16]. Reaction time in batch processes was considered as an independent variable within the experimental design in some cases and in others it was kept out of the Box-Wilson design and fixed under the experimental criteria. Finally, the grafting extension was determined by an improved hot titration method [11, 13, 14]. Additionally, infrared spectra were recorded and the presence of grafted SA, SF or *p*PBM groups was determined [11-16]. Heterogeneity of grafts was determined by *n*-heptane selective extraction. Further characterisation was done, among others, by synchrotron wide-angle x-ray scattering and small-angle x-ray scattering techniques, thermal analysis, optical microscopy, and so on. For more detailed information see the references included at the end of the chapter.

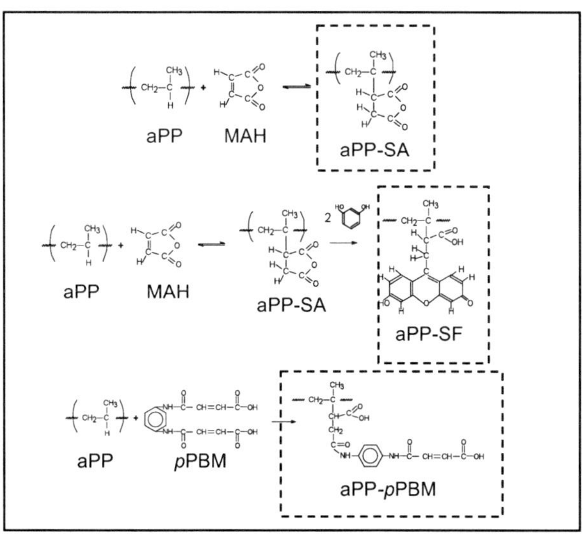

Figure 3.1 Reaction schemes and grafted species

Despite the huge efforts conducted in the study of the chemical modification process in the last 40 years, the real nature of the processes of chemical modification of polyolefins, mainly PP, is still to be clearly ascertained. In fact, nowadays there still exists a great controversy about the mechanisms governing the process. The latter is partially caused, by the difficulty of quantifying accurately the degree of grafting and conversion of the modified polymers. In fact, when consulting literature, it is observed that the terms for degree of grafting and for conversion are often misleading. Also, there are a great amount of data in literature that exhibit a lack of robustness that makes it very difficult for them to be used in proper mechanistic studies. In fact, an extended opinion in academia is that the chemical modification of polyolefins throughout is chaotic and results are non reproducible. However, authors are not in agreement with this opinion, sometimes based on isolated non-robust experimental data making difficult further interpretations about the reaction mechanism, nevertheless if experimental research is supported by statistical methods the resolution of the problem goes ahead. The latter appears as one of the reasons why we undertook these type of studies by embracing Statistical Design of Experiments (SDoE), in particular Box-Wilson experimental procedures, in order to provide of robustness to laboratory data as well as to make sure that the system studied is sufficiently precise and that the results obtained are reliable enough, at least in the experimental co-ordinates space considered. Moreover, because of the role played by the reaction time all along the process, the complex nature of the process of chemical modification of polyolefins by polar monomers (such as those proposed here) look to justify the high level of efforts devoted to the topic as can be deduced for the great amount of reviews in the literature related to the topic [17-21].

For all the reasons mentioned previously, although grafting processes have been widely studied, the real chemical pathways involved are still far from being understood as there exists a great controversy in the literature about the role of the species involved in the process as well as the preferential location of the grafted moieties onto the polymer backbone. Nevertheless, the existence of competitive reactions in the polymer bulk is generally accepted. These lead to in-chain scission or the degradation of PP. It is perhaps well worth mentioning that the use of Box-Wilson experimental designs has provided a tool to not only to follow the evolution of the grafted groups and conversion in the experimental co-ordinates space considered but also to obtain models capable of forecasting the behaviour of the system in the correct manner. For these kind of experimental designs the authors have tried to mimic Nature by considering that a chemical modification process resembles a prey-predator system in a defined ecological niche. So, since the behaviour of these type of systems in Nature are usually successfully modelled by quadratic equations that include the factors affecting the media, the choice of using Box-Wilson quadratic models appears as very suggestive under the described circumstances and bearing in mind that the factors in the experimental design has been carefully chosen under the consideration of being the prime affecting the overall behaviour of the modification process. Such circumstances allowed the authors to define a reaction pathway [15, 16] that provided a strong explanation as to what is happening in the chemical modification process for whatever the reaction media used for modifying the polymer. **Figure 3.2** shows the species involved in the chemical modification process performed in solution (a) and in the melt (b) after balancing the species involved. From these reaction schemes the fact that the reaction yield is higher in the melt than in solution (if performed under the same reactant concentration and reaction time) clearly emerges. The latter definitively finish with the controversy about the preference of solution processes looking for a more homogeneous reaction media, good stirring, better temperature control, and so on, respecting the molten state process [15, 16]. At a glance, the mechanism in **Figure 3.2** allows the reader to ascertain that to graft two SA groups onto the polymer chain requires three molecules of both MAH and initiator (RH) in the case of the process in the melt. Meanwhile, if performed in solution, to obtain the same number of grafts would require twice the amount of MAH and RH. However, the explanation of the process is far from being so simple. The comparison of the one performed in solution or in the melt implies a lot of additional factors such as degradation possibilities, radical activity transfer, preferential location of grafts, and so on. Only to mention here that these aspects are outside of the scope of this chapter, so, the reader interested in all the mechanistic aspects of the chemical modification process is directed to references about [15, 16]. From the analysis of the species involved in **Figure 3.2**, jointly with the robust data obtained for the Box-Wilson worksheet, the authors proposed an unsteady state process mechanism that agrees with the fact (already observed in previous works on the isotactic polymer) [22] that short reaction times were not only sufficient but necessary to obtain the highest possible number of grafted species while preventing degradation processes [15, 16]. This fact appears to explain why reactive extrusion has proved to be an efficient method for obtaining a great part of the interfacial agents' portfolio in the market. Otherwise, it is well worth remembering that the reaction media greatly condition the architecture of the modified polymer. The latter is mainly by the inter- and intra-chain radical activity transfer possibilities as well as the different ways for the termination reactions to evolve. So, and according to reaction schemes in **Figure 3.2**, the processes performed in the melt would yield a higher degree of grafting. This one is what exactly occurs. For example, when the processes are performed using similar reaction times and reactant concentration conditions as published elsewhere [15, 16], the degrees of grafting obtained were in the 0.5% up to 2.5% interval for the aPP modified in solution, and in the 0.5% up to 4.2% for those aPP-SA obtained in the melt. The higher degree of grafting on aPP modified in melt state than those modified in solution processes apply also in the case of the obtention procedures for aPP-SA and aPP-*p*PBM that even exhibit the same values if grafting values for both monomers, MAH and pPBM acid, are expressed in SA moieties equivalent.

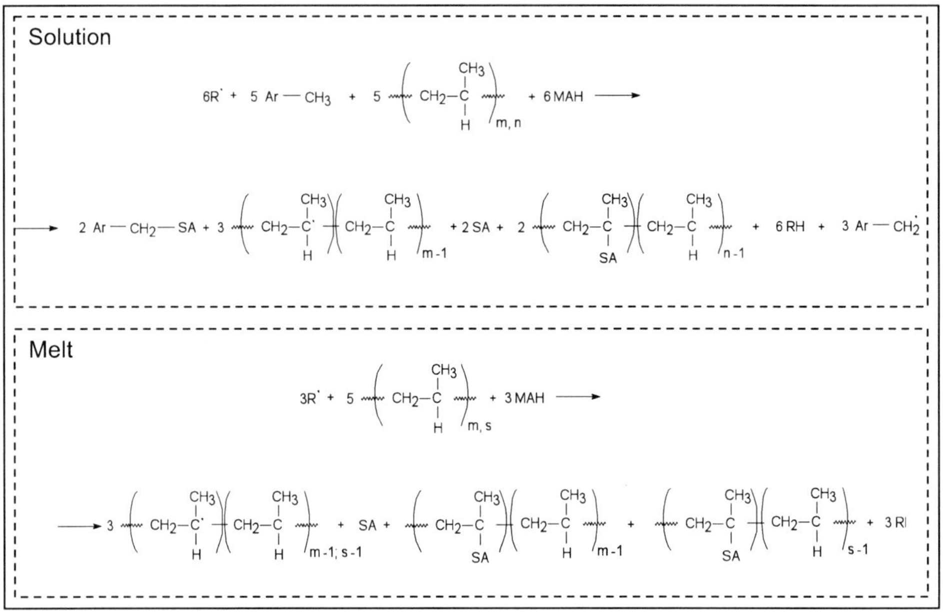

Figure 3.2 Species involved in the chemical modification process

As it has been mentioned previously, the reaction media highly condition the architecture of the grafted polymers such as is shown in the consecutive reaction scheme in **Figure 3.3** and proposed elsewhere [7, 8].

In fact, **Figure 3.3** shows the typical pattern corresponding to a consecutive reaction scheme where the desired product must be obtained at a critical reaction time afterwards a sharp decrease in the yield of the target product and an increase of other non desired (or less valuable) products occur. Merely as an example of the latter, **Figure 3.4** shows the evolution of grafting with reaction time for a fixed amount of reactants either in solution (MAH) or in the melt (*p*PBM), and where the observed evolution agrees with the pattern showed in **Figure 3.3**.

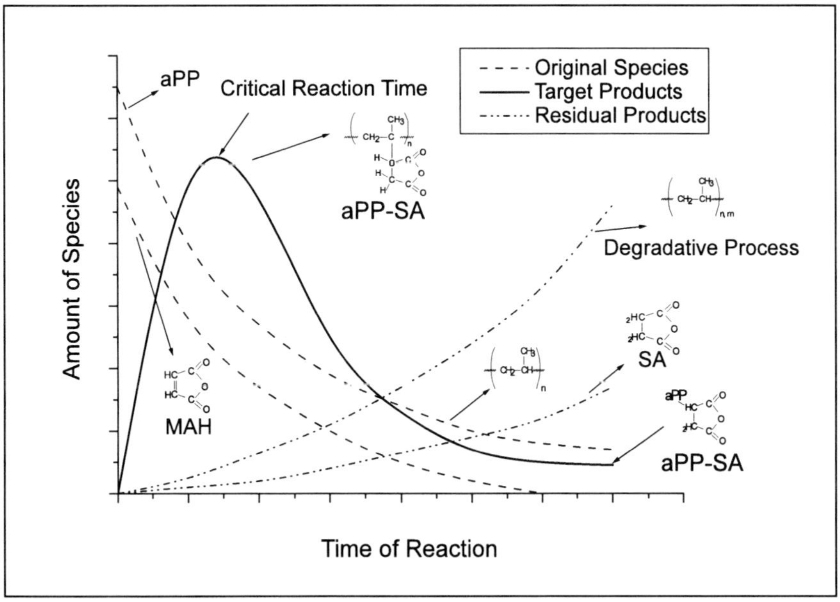

Figure 3.3 Consecutive reaction scheme

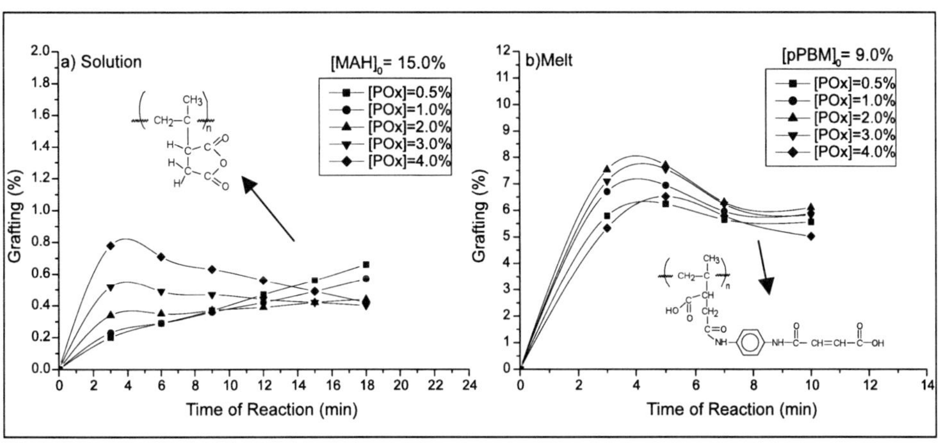

Figure 3.4 Grafting *versus* reaction time for the (a) solution, and (b) melt conditions showed at each case

Indeed, by observing this **Figure 3.4** we can realise that meanwhile the process performed in solution the higher initiator (at a fixed amount of initial polar monomer) the higher the grafting level, the process in the melt is showing an optimal yield not at the highest level of initiator but at an intermediate value of 3%. So, in the melt after surpassing a critical level in initiator amount, increasing amounts of it imply a lower conversion of grafted polymer. The latter is strongly related to the capacity of transferring radical activity to the reaction media. A more profound explanation of this phenomenon can be read elsewhere [5, 15, 16]. Besides, the latter agrees with the fact that the process in the melt through grafting values much higher than in solution. For example, the highest grafting level in **Figure 3.3a** (aPP-SA) yield a value close to 0.9% (0.9x 10^{-4} mol/g), meanwhile that in **Figure 3.3b** shows a value close to 8.5% (2.8x 10^{-4} mol/g), almost three times higher (by considering the molecular mass of the different grafted group) than in solution.

Looking to obtain a new type of grafted PP, the condensation reaction between resorcine molecules and grafted SA groups on the aPP (**Figure 3.2**) has allowed the authors to establish that SA grafts are monomer units rather than polysuccinic units as early studies proposed. The fact that in solution the condensation reaction leads to grafted polymers with both SA and SF groups due to the SA moieties forming an anhydride bridge between two PP sequences makes it unable to be reacted by resorcine. So, it can be concluded that in the melt, the termination processes evolve to disproportionation yielding the grafted group as an end group while in solution a great part of the processes does by recombination yielding SA groups grafted side to side to PP chains. Even more, aPP has allowed conclusions about the role of other effects in the reaction yielding (stereoregularity, temperature, kind of solvents, and so on). Hence, because of their different molecular architectures depending on the kind and the reaction conditions of their obtention processes, the interfacial agents obtained may play very different roles as interfacial modifiers in composites and polyblends as described elsewhere [9-14].

3.3 Role in Heterogeneous Materials Based on Polymers

Although the main subject of the present chapter is to show the way in which an industrial by-product (or waste) is converted into another material with very different properties itself but which 'high value' comes from its ability of improve the performance of heterogeneous systems based on polymers, the following section shows a brief overview of the strategy followed in these type of studies. So, once the interfacial modifiers from the matrix side were available they were tested in chosen samples and contrasted with the performance offered by commercial contra-types based on isotactic polypropylenes [11]. Due to the very good results obtained in these preliminary stages the following step has been to undertake the modelling of the behaviour in terms of SDoE approaches (such as Box-Wilson) by considering this kind of system as agent-based models [10]. A selection of the interfacial agents obtained from the aPP by-product has been evaluated in composite materials based on iPP matrices incorporating talc, mica and carbon fibres (both sized and nanofibres grown from the different organic vapour phases), and the efficiency of them has been ascertained. In this case, the proper action of the functionalised PP can be associated to the increase of the second order interactions between the matrix and the dispersed phase as well as to the higher mobility of polar groups bonded at the chain ends, jointly to its ability to be hosted at the interphase between the reinforcement and the matrix, especially useful to reduce the interfacial tension between the components of the heterogeneous system. The presence of the reinforcements such as talc or mica, as odd solid particles embedded into a PP matrix, usually induces a nucleation effect evidenced by a significant increase in the crystalline content of the polymer if compared with the neat polymer when processed at the same setup conditions that are necessary to ensure a good accommodation of the solid particles into the amorphous phase of the polymer in order to obtain a material with a good mechanical performance. In fact, it has been found that there is a very good agreement between the mechanical properties obtained and the morphology of the composites. In these studies, since the interfacial regions must be considered, it becomes necessary to differentiate between the amorphous/crystal polymer interface and the amorphous polymer/mineral interface where the interfacial agent trends to be embedded [5-7]. Dynamic mechanical analysis (DMA) measurements mainly related to the relationship between the heterogeneous material mechanical behavior and the relaxation spectra of the polymer matrices also have proved to be very useful to discuss the role played by the interfacial agents depending on their amorphous or semicrystalline character [5-7]. This is the kind of remarks to consider about the finite dimensions of the interfacial area at the polymer/mineral interphase when trying to modify it by replacing a part (frequently just a little part is not only necessary but sufficient) of the PP matrix by the interfacial agent. The latter gives an idea why the interfacial agents derived from aPP works properly in these kind of systems. As an example **Figure 3.5** shows the variations of the elastic modulus of the iPP/interfacial modifier/talc composites as a function of either the type of grafted group: SA or SF, as the degree of grafting and the amount of interfacial agent present. A comparison to a commercial additive is also included [5-7]. It is important to note (from data in **Figure 3.5**), that by maintaining as constant the amount of talc and mica (30% *w/w* in this example), it is possible to obtain materials with a very different stiffness (values between 2 and 5 GPa) by varying only the type and the grafting level of the interfacial agent with a mere 1.5% of this in the polymer phase of the composite. Also, it is important to note that the ultimate behaviour of all these composites (modified with the indicated interfacial agents obtained from the polymer wastes) appears as comparable (and even superior) to those with the most efficient (in our experience) of the commercial interfacial agents [11, 23]. The possibility of obtaining such a broad spectra of properties gives a real chance of obtaining the so called 'tailor made' materials. That means, to design materials with previously defined properties and/or performance, design of materials rather than with materials. Of course, the other parameters associated with the tensile properties (and mechanical properties in general) also exhibit a similar evolution [5-9].

According the 'Top-Down' approach followed by authors in the experimental planning design, once the macroscopic responses of the material have been optimised because of the interfacial modifications induced in the cited polymer composites, they have also been detected for whatever the techniques purchased such as mechanical properties (tensile and flexural conditions), thermal differential scanning calorimetry (DSC), dynamic mechanical analysis (DMA), optical microscopy and thermo-optical microscopy, scanning and field emission electronic microscopies, and synchrotron Fourier-transform infra-red spectroscopy. On such an approach is based the modelling of the material's performance.

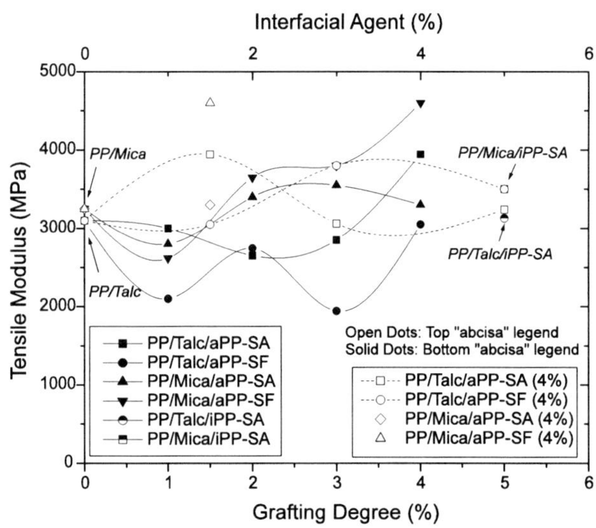

Figure 3.5 Evolution of modulus with the interfacial agent amount and grafting degree of the indicated samples

Indeed, as has been mentioned before, DMA spectroscopy has proven to be very useful in the study of the mechanical properties of a material in dependence with temperature, implying the possibility to follow the type of relaxation phenomena observed. The possibility of discussing separately the elastic behaviour of the composite (storage modulus) and the viscous response of the same (loss modulus) offers plenty of possibilities for a fine interpretation of the interfacial phenomena. Since an exhaustive discussion of these aspects is far from the purpose of the present chapter, the reader can find more information about the composite systems studied by authors in articles elsewhere [5-7, 9]. As an example **Figure 3.6** shows the evolution of both the storage modulus (E′) and the loss modulus (E″) with the temperature in the -30 °C up to 150 °C range. So, the effect of the interfacial agent (in this case aPP-*p*PBM) by considering the temperature dependent relaxation phenomena can be ascertained. **Figure 3.6** shows the behaviour of iPP (neat PP) and two composites with 25% *w/w* of mica (with and without interfacial agent). So, **Figure 3.6** (solid dots) shows how the presence of the mineral increases the value of E′ in all the temperature ranges, in accordance with the nucleating effect of the mica. Furthermore, it can be observed that a little amount of aPP-*p*PBM implies a high improvement in the elastic response of the system that confirms the very efficient role played by this interfacial agent (aPP-*p*PBM) obtained from polymer wastes. **Figure 3.6** (open dots) shows how the viscous response of the system is higher for the unmodified PP/mica system. This is related to the fact that the amorphous phase (where the mineral is necessarily embedded) is highly constrained in the unmodified system while by the presence of aPP-*p*PBM at the mineral/matrix interphase the viscous component (E″) not only decreases all along the whole temperature range showing the same pattern than the neat PP, but in this case even a more amount of amorphous phase at the amorphous/crystal interphase of the matrix and the amorphous bulk of the material is 'free' to participate in the cooperative segmental motions considered as the responsible of the T_g of the polymer matrix. The increase on the peak area under this transition region at **Figure 3.6** (open dots) is supporting this fact.

In the case of heterogeneous materials (based on organic polymers) with a deformable dispersed phase, it means, polymer blends and alloys, the interfacial activity of the previously mentioned interfacial agents has been also evaluated by checking the compatibilisation capability of two immiscible semi-crystalline polymers such as iPP and polyamide 6 (PA). From an academic viewpoint, this family of blends are an excellent model for most general polycondensation/polyolefin polymeric systems by focussing on the development and control of post-reactive processing stable morphologies. When optimised, these tailor made blends are characterised by good processability, low water absorption, low liquid and vapour permeabilities, improved dimensional stability, good impact strength and improved chemical resistance to alcohol and glycols. The study of this system by means of Box-Wilson experimental designs has allowed the proposal of a micromechanical model based on the possibility of chemical reaction through the interphase in dependence of the type of polar group of the interfacial agent [8], that has been recently validated by means of field-emission scanning electron microscopy (FESEM) on tensile test fractured samples. Nevertheless, the evaluation of the interfacial agents obtained from aPP in this system is running nowadays, although studies on mechanical and thermal behaviour have already been published [4, 8] and the capability of these interfacial agents (mainly aPP-SA) of generating morphologies stable even after a post-processing step has been demonstrated. Finally, and as a mere example of the action of the interfacial agents only to show the differences in morphology (in even the most difficult scenario: a 50/50 blend merely compression moulded processed) of a PP/PA 50/50 sample without and with modified interphase by the action of 9% aPP-SA (over the PP phase) obtained from industrial wastes (**Figure 3.7**).

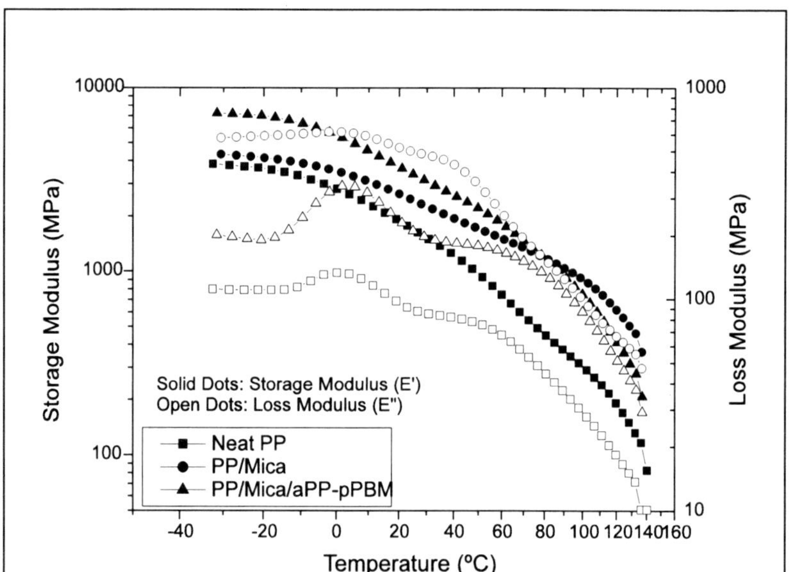

Figure 3.6 Evolution of storage modulus (E´) and loss modulus (E´´) with temperature for a number of compounds without and with aPP-*p*PBM as interfacial modifier from the matrix side

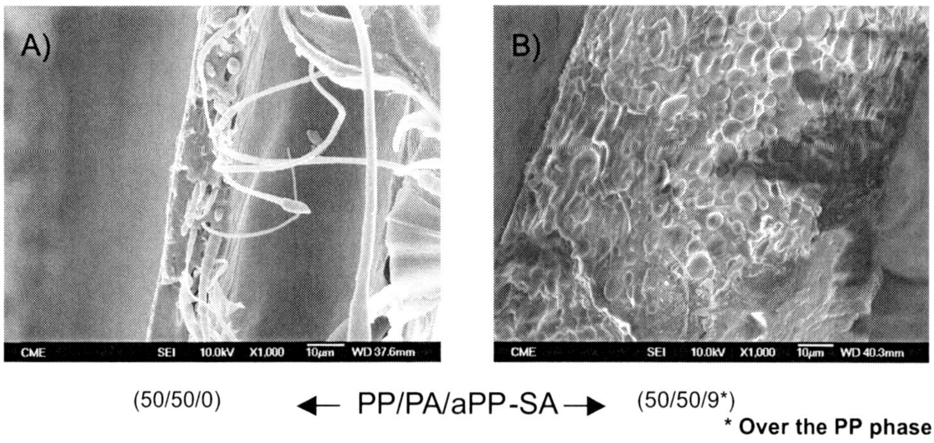

Figure 3.7 FESEM images (1000x) of a PP/PA 50/50 blend without aPP-SA (A) and with aPP-SA (B)

3.4 Conclusions and Perspectives

The necessity of obtaining proper interfacial modifiers (or agents) from an industrial waste have been proven. The ability of obtaining modified polymers with different functionalities (as desired) jointly to the future perspectives of being able to allocate preferentially the grafts allow us to use them as interfacial materials for whatever performance required between the limits imposed by the matrix in the heterogeneous material. The analysis of data aided by SDoE approaches open a wide window in the concerning to tailor made materials. The fact appearing to emerge is that the experimental strategy followed allows obtaining the modified polymer with the desired number of grafts by dosing conveniently the reactants and controlling the reaction conditions according the Box-Wilson polynomial forecasts. So, the final goal would truly be a control over the molecular architecture of the chemically modified polyolefins and how this fact should affect their efficiency as interfacial modifiers in heterogeneous materials based on organic polymers.

Hence, although serendipity has played a prime role in the ultimate efficiency of these interfacial agents, they will be the markets needs which will define the ultimate function of the target material. Moreover, as it has been recently written, the story of polyolefins can be considered as the quintessential example of the convergence of creativity and practically leading to the commercial launch of new materials [24], consequently the development of those materials based on them would follow the same path. Nevertheless, since novel polymeric materials (and their derivatives), are sought to replace existing ones, the success of the new ones will be strongly dependent on the markets motivation of change, combined with the sustainable growing and environmental engagements. It means that the novelties must offer a superior performance jointly with advantageous savings and cost reductions [24].

References

1. D. Tripathi, *Practical Guide to Polypropylene*, Rapra Technology Ltd., Shrewsbury, UK, 2002.

2. J.A. Brydson, *Plastics Materials*, 7th Edition, Butterworth-Heinemann, Oxford, UK, 1999.

3. E.P. Moore, *The Rebirth of Polypropylene: Supported Catalysts: How the People of the Montedison Laboratories Revolutionised the PP Industry*, Hanser, Munich, Germany, 1998.

4. G. Natta, F. Beati and F. Severini, *Journal of Polymer Science*, 1959, **34**, 685.

5. J.M. García-Martínez, S. Areso, J. Taranco and E.P. Collar in *Polyolefin Blends*, Eds., D. Nwabunma and T. Kyu, Wiley and Sons, Hoboken, NJ, USA, 2008, p.379.

6. J.M. García-Martínez, O. Laguna, S. Areso and E.P. Collar, *Journal of Polymer Science Polymer Physics*, 2000, **38**, 1554.

7. J.M. García-Martínez, O. Laguna, S. Areso and E.P. Collar, *Journal of Applied Polymer Science*, 2001, **81**, 625.

8. J.M. García-Martínez, S. Areso and E.P. Collar, *Journal of Macromolecular Science, Part B: Physics*, 2001, **B40**, 387.

9. J.M. García-Martínez, O. Laguna, S. Areso and E.P. Collar, *European Polymer Journal*, 2002, **38**, 1583.

10. J.M. García-Martínez, S. Areso and E.P. Collar, *Journal of Applied Polymer Science*, 2009, **113**, 3929.

11. J.M. García-Martínez, O. Laguna and E.P. Collar, *Journal of Applied Polymer Science*, 1997, **65**, 1333.

12. J.M. García-Martínez, O. Laguna and E.P. Collar, *Journal of Applied Polymer Science*, 1998, **68**, 483.

13. J.M. García-Martínez, O. Laguna, S. Areso and E.P. Collar, *Journal of Applied Polymer Science*, 1998, **70**, 689.

14. J.M. García-Martínez, A.G. Cofrades, S. Areso and E.P. Collar, *Journal of Applied Polymer Science*, 2003, **88**, 2202.

15. J.M. García-Martínez, S. Areso and E.P. Collar, *Journal of Applied Polymer Science*, 2006, **102**, 1182.

16. J.M. García-Martínez, S. Areso and E.P. Collar, *Journal of Applied Polymer Science*, 2007, **104**, 345.

17. G. Xu and S.J. Lin, *Journal of Macromolecular Science - Reviews in Macromolecular Chemistry & Physics*, 1994, **C34**, 555.

18. Y.H.R. Jois and J.B. Harrison, *Journal of Macromolecular Science - Reviews in Macromolecular Chemistry & Physics*, 1996, **C36**, 433.

19. M.K. Naki and M.S. Choudhari, *Journal of Macromolecular Science - Reviews in Macromolecular Chemistry & Physics*, 1996, **C36**, 601.

20. G. Moad, *Progress in Polymer Science*, 1999, **22**, 81.

21. E. Passaglia, S. Coiai and S. Augier, *Progress in Polymer Science*, 2009, **34**, 911.

22. J.M. García-Martínez, J. Taranco, O. Laguna and E.P. Collar, *International Polymer Processing*, 1994, **9**, 246.

23. J.M. Felix and P. Gatenholm, *Journal of Applied Polymer Science*, 1991, **42**, 609.

24. P.D. Hustad, *Science*, 2009, **325**, 702.

4 Energy Efficiency Index for Plastic Processing Machines

Rainer Pamminger, Erik Sehnal and Wolfgang Wimmer

4.1 Introduction

Due to the growing global demand for energy and raw materials and due to their limitations, the issue of resource efficiency and energy efficiency becomes increasingly important. Companies that want to be competitive in the future have to handle the available resources carefully.

Plastic processing machines like injection moulding or extrusion machines are energy intensive products. These machines are running on a two or three shift basis and have a significant energy consumption. Most of the environmental impacts occur during the use phase. In comparison to the other life cycle phases, the use phase shows the most relevant environmental impact. Therefore the other product life cycle phases are in most cases less important and the improvement process should focus on the use phase.

For the comparison of the environmental impact of the machines, the energy consumption has been chosen. Considering carbon dioxide emissions in this case would not lead to meaningful results, as it is highly dependent on the energy mix in the country where the machine is running. Therefore, the carbon footprint, is only meaningful for a whole site but not for an individual machine.

Results from a previous project show that for plastic processing machines, 87% of the environmental impacts are due to the energy consumption during the use stage (**Figure 4.1**).

In the past, machines were very difficult to compare their energy efficiency. Depending on the machine type, the plastic used, and the processing cycle the energy consumption varies widely. Additionally, due to a large variety of products with large differences in processed weight (from a few grams up to several 100 kg) and various forms (individual for each application), it is difficult to define a comparable energy consumption.

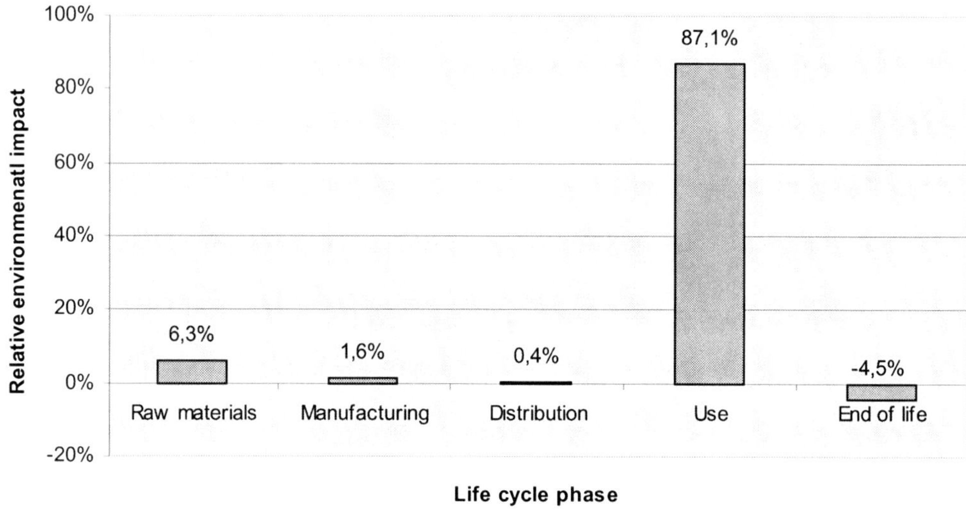

Figure 4.1 Environmental profile of an injection moulding machine [1]

For injection moulding machines an energy measuring standard, the EUROMAP 60 [2] (European committee of the national associations of machinery manufacturers for the plastics and rubber industries), already exists, but according to Vettkötter [3], the standard has been published when the electrical drive technology was not so relevant then as today. A survey of European manufacturers showed that the measurement standard is more or less not in use.

4.2 Aim and Benefits of the Energy Efficiency Label

The main goal of the energy efficiency label was to reduce the energy consumption of plastic processing machines. Labels shift the distribution of energy efficient models upward by providing information that allows consumers to make rational decisions and by stimulating manufacturers to design products that achieve higher ratings than the minimum standard [4].

A good example is the European Union (EU) label for refrigerators which was introduced in 1996. In **Figure 4.2** the product sales in Germany assigned to energy label categories from 2000 to 2007 are displayed. The label has led to a rise in energy efficiency from 1996 to 2000 by over 30% [5] and between 2000 and 2007 also to a similar improvement in energy efficiency.

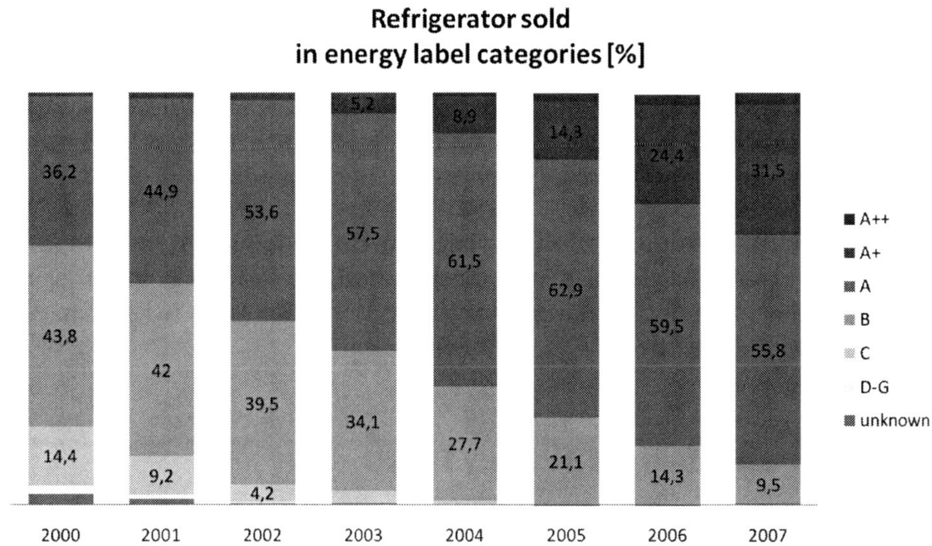

Figure 4.2 Product sales assigned to energy label category from 2000 to 2007. Reproduced with permission from GfK Panelmarket, Verkaufte Kühlgeräte nach Effizienzklassen in den Jahren 2000 bis 2007, Nuremberg, Germany. 2008. ©2008 GfK Panelmarket [6]

The introduction of an energy label and the reduction in energy consumption obtained has a positive impact on *manufacturers* and *customers* in the plastics industry as well as the *environment*.

From the *customer* point of view the label allows an objective comparison of energy consumption of different manufacturers. Because of the energy measurement standard the energy consumption has to be measured in the same way every time and is therefore comparable.

This leads to more thoroughly investigated and reliable decisions with the main focus on the energy demand and therefore also on the energy costs at the use phase.

Implementing an energy efficiency label as a communication tool will strengthen the demand for energy-efficient machines. Consumers will be much more aware of the importance of the energy consumption of the machines and will use it as a purchasing criterion. The energy consumption of the user will be reduced and therefore the energy costs.

For the *manufacturer* it them the possibility of advertising that they are producing high efficiency plastic processing machines. Using this label will give companies with energy efficient machines a competitive advantage in the market. Other manufacturers are then highly motivated in developing energy efficient machines themselves. For this improvement process it is also very helpful to support the producers with possible ways on how to improve the energy performance of the product.

Using this communication tool will strengthen the demand for energy-efficient machines and reduce the global energy consumption. This can help countries meet their climate change goals. Reducing electricity consumption leads to less demand of power plants and decreases the CO_2 emissions.

4.3 Definition of Energy Efficiency Labels

Energy efficiency labels are informative labels allocated to a manufactured product to describe their energy performance. These labels give consumers the data necessary to make energy efficiency a factor in their purchasing decisions [4].

Generally labels can differ in many ways: mandatory and voluntary labels, endorsement and the comparative labels, single use or multiple use labels, and so on.

Mandatory labels have to be applied by law, for example, EU labels for refrigerators and voluntary labels can help best in class products to communicate their performance voluntarily on the market.

Endorsement labels can also seen as 'seals of approval' where specified criteria are fulfilled [4]. Comparative labels give consumers the opportunity of comparing the performance of similar products using either discrete categories of performance or a continuous scale. In this chapter the main focus is on comparative labels, because labelling the energy efficiency of industrial machines by means of endorsement labels is not as meaningful.

The energy efficiency labels also can be classified by the credibility provided and the number of possible users of the label (**Figure 4.3**).

The *credibility* depend on how much information on the measuring method and the efficiency criteria is provided, for example, the energy label for pumps has a high credibility, as the measuring method, the efficiency criteria and the energy classification can be downloaded on their webpage (*http://www.europump.eu*). Arburg's seal of approval 'e^2' (*http://www.arburg.com*) is a label where the criteria used and the measurement standards are not that clearly communicated.

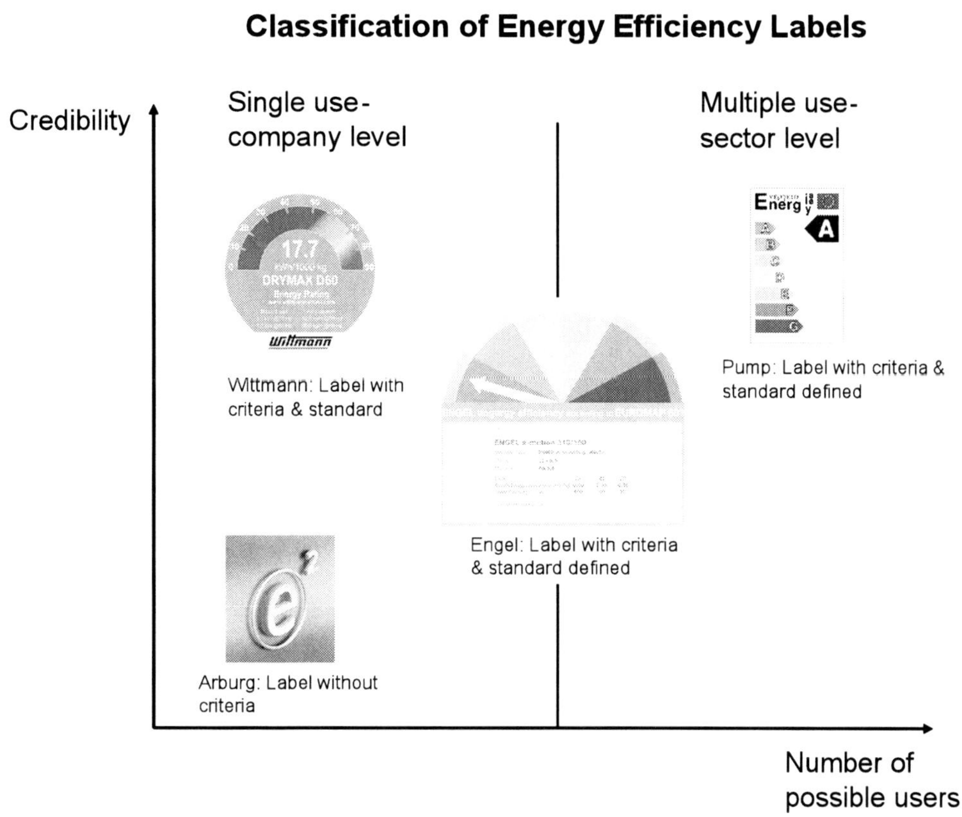

Figure 4.3 Classification of energy efficiency labels

The **number of possible users** describes the companies who could use the label. For example the company Wittmann has developed their own label for dry air dryers [7]. As it is a company label - single use - no other company can use it and therefore a comparison is difficult. If the label was for multiple use, e.g., the criteria are available for the whole branch, this gives a high credibility. The pump industry, for example, developed an energy label, voluntary for all manufacturers but frequently used due to the market pressure.

The injection moulding company Engel also recently developed an energy efficiency label (see **Figure 4.3**) using the newly developed energy measuring standard for injection moulding machines EUROMAP 60 [8]. With the EUROMAP 60 and the defined measuring method this label has a high credibility and the standard is open for the whole injection moulding branch. The label design itself is the company's own and can be used by Engel only. In addition to the results belonging to the EUROMAP 60 the label gives a graph indicating the energy efficiency of the machine. The classification and the grading in this case is unclear and has low credibility.

4.4 Label Development Process

The procedure of developing an energy efficiency label will include the steps shown in **Figure 4.4**.

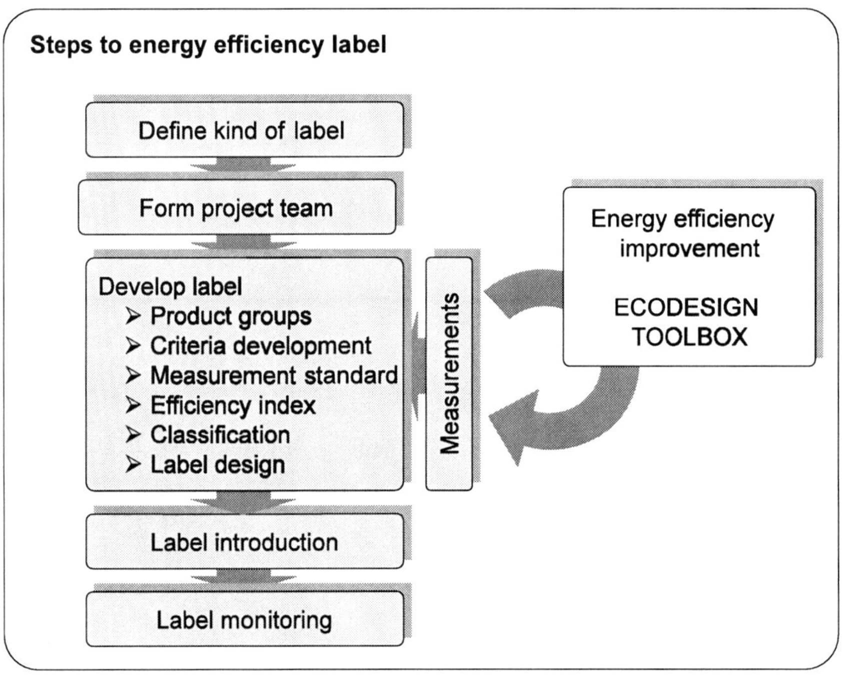

Figure 4.4 Steps creating an energy efficiency label

4.4.1 Define the Kind of Label: Which Type of Label do we need?

In this step it has to be clarified whether the label should be a seal of approval or comparative. The seal of approval gives just the information that some criteria are fulfilled but not to which level and therefore no comparison is possible with the comparative label. Then the credibility and the possible number of users have to be defined at the beginning of the label development. Additionally it should be decided to focus on a voluntary or a mandatory label.

4.4.2 Form a Project Team: Who should be Involved in the Label Development Process? Which Steps have to be Done and When?

Depending on the label-type the project team has to be defined in different ways. For a single use – company label, the team should include persons from product development, measurement engineering and marketing. If a label for multiple use is to be developed, the project team should include the industry representation body (if one exists) and representatives of all well-known companies. If the label should be mandatory for all companies in the branch, a legislative and administrative authority must be considered to conduct the label development steps and to decide about the new standards needed. For voluntary labels this will be done by an implementing body and should be a transparent process.

4.4.3 Definition of the Product Groups: Which Product Groups/Segments can be Defined and Considered Together?

In this step the product types covered by the label have been chosen and grouped. For choosing the product types the market share and the environmental impact can be used. Grouping of products is necessary as not every product will have its own energy measurement standard. Here the technical challenge is to group products with similar technical characteristics and similar energy values. For each group a specific energy measurement standard has to be created.

4.4.4 Definition of Criteria: Which Efficiency Criterion can be used for the Evaluation of the Energy Efficiency?

For the evaluation of the energy efficiency qualitative and quantitative criteria are possible. Qualitative criteria such as use of certain components are easy to check and common options for seals of approval. Quantitative criteria such as the measured specific energy consumption (SEC) in kWh/kg require a measurement standard and measuring procedures. Such quantitative criteria are more significant and allow direct comparison, but are more difficult to obtain and control.

4.4.5 Developing Measurement Standards: How to Measure the Energy Consumption of the Product?

A common measuring method for the defined quantitative criteria is critical for the development and introduction of the label of the product. All manufacturers have to be evaluated in the same way. In this measuring standard the procedure, system boundaries, standard test facilities, and so on, have to be defined.

4.4.6 Calculate the Energy Efficiency Index (EEI): How to Define an EEI?

For each product category an EEI has to be defined. Therefore, a formula pointing out the energy efficiency of the product has to be defined. The value of the EEI should describe the energy efficiency in a reasonable way for example 40%. This EEI can be defined as a ratio of the theoretical energy demand needed for melting the plastics and the actual energy demand from the energy measurements according to the measurement standard.

4.4.7 Classification of Energy Classes: How Can Products be Classified?

Label classes are a typical way to show the level of energy efficiency of different products. One example of many is the efficiency classes of the EU label for washing machines (classes A-F). For such classifications, the EEI values have to be classified. The range of EEI in each class can be easy defined according to best available technology and best not available technology (BNAT).

4.4.8 Label Design: How the Label is Designed and which Information is Included?

According to the requirements of the customer and to the measuring standard the needed data is illustrated on the label.

4.4.9 Energy Measurements: How to Provide Data for the Definition of the Measurement Standard and the Definition of the Energy Classes?

Parallel to the label development energy measurements have to be carried out. This helps to define and prove the measurement standard. The results of the energy consumption and the derived EEI are then used for the definition of the energy classes.

These energy measurements are also a starting point for the improvement process and an assessment of the technology potentials.

4.4.10 Energy Efficiency Improvement: What are Possible Improvement Strategies for a Higher Energy Class?

For improving the efficiency of the product and receiving a higher energy class ranking the manufacturer can be supported in the product improvement process. For example the 'Ecodesign Toolbox' [9] is a seven step approach for the environmental improvement of products.

4.4.11 Label Introduction

The effectiveness of energy labels is highly dependent on how they present information to the customer and on how they are supported by marketing and information campaigns. Consumer confidence and acceptance in the label and its reliable indication of energy efficiency is the key to establish an energy efficiency label that can affect consumer behaviour towards an investment decision considering energy efficiency.

4.4.12 Label Monitoring

Once the label is introduced, the labelling has to be monitored in terms of reliability and accuracy information. This can be realised by mandatory labelling schemes and procedures or for voluntary labelling programs by self-regulation schemes, among the competitors using the label. The manufacturers will declare their products themselves based on the defined label regulation. The accuracy and reliability of the data is competitor-controlled.

4.5 Example: Plastic Extrusion Machines

In the following sections, the steps to derive an energy efficiency label for plastic extrusion machines as performed in the project '*Development of criteria for the communication of the energy efficiency of plastic processing machines*' [10] are shown.

4.5.1 Label Definition and Project Team

The label was aimed as a multiple use label for a defined segment of the extrusion market. The concept of the label developed in a research project should be further accomplished and introduced on the European marked from the manufacturers representatives such as the VDMA or EUROMAP.

4.5.2 Label Development

The label was defined as part of an energy efficiency project involving manufacturers, plastics processors and research companies.

The energy measurement standards was discussed and proposed in the EUROMAP 30 [11] to establish a market-wide accepted standard. This process is still ongoing.

4.5.3 Energy Efficiency Criteria

For the criteria a quantitative efficiency criteria standard was chosen, utilising the comparable and established indices of the SEC in Wh/kg. In this index, the energy consumption is referenced to the amount of commodity produced in kg. The output is measured on a regular basis and reported in kg/h. The abbreviation SEC is normally used:

$$SEC\,[Wh/kg] = \left(\frac{Energyconsumption\,[kWh]}{Output\,[kg]}\right) \times 1000$$

There are no restrictions on how to achieve energy efficiency. This allows the manufacturer to develop innovative technological improvements with the focus on low energy consumption.

The system boundaries, boundary conditions and duration of the measurement have to be defined. A comprehensive measuring standard is aspired that defines a norm effective in an industry sector for the determination of the energy consumption for the label.

The use of different forms of energy can pose a big problem when determining the specific energy consumption, i.e., if in addition to electricity other forms of energy are also used such as gas or heat. In order compare and evaluate them, comparable parameters have to be defined that allow assessment of the primary energy consumption. As an alternative, the label could also show the different kinds of energy used.

4.5.4 Energy Measurement and Measurement Standard

In the plastics processing industry different types of machines often cannot be compared properly due to the large variety of applications and products. When developing a measuring standard, it is most important to use a scenario that is as near to the real industrial use of the machine as possible in order to determine the actual energy consumption. Due to the large variety of methods, applications, and products the consistent definition of measuring standards is therefore very difficult but of utmost importance.

In this case study, the categories polyvinyl chloride (PVC) profile extrusion and polyethylene pipe extrusion have been selected. Within these sectors, there are comparable products on the market and considerable market sizes exist. The measurements are taken at a machine producing with typical operating conditions of this market segment.

Therefore the material (PVC-dryblend), the mass pressure (30 ± 1 MPa) and mass temperature are defined. The measurements can be done with a test tool or with a test nozzle.

Within two cycles the effective power (kW) and the output (kg/h) have to be measured. One cycle is in full load (100% of the guaranteed output) and the second is in partial mode (60% of the guaranteed output). The results are then combined to obtain an overall figure.

The integration of the industrial agencies EUROMAP representing the most important companies in the market is a very important factor for the worldwide propagation and recognition of the standard. The evaluation of the measurement standard is currently ongoing.

A label for an industry sector will only be successful if it is based on a measuring standard that is accepted and comprehensible inside the community.

4.5.5 Energy Efficiency Index (EEI)

As a next step a table of indices is developed based on measurements, best practice research and estimations of the potential to increase energy efficiency. This table poses as a foundation to the energy efficiency label. Subsequently the EEI can be classified.

It is reasonable to define an EEI in order to do a classification. It is calculated from the measured SEC according to the measuring standard, the amount of energy needed theoretically to melt a certain polymer and, in some cases, constant correction factors:

$$EEI = \left(\frac{E_{theor}}{E_{act}}\right) \times X_{korr} \times 100 \text{ in \%}$$

E_{theor} Amount of energy needed theoretically to melt the material (kWh/kg)

E_{act} Measured SEC according to measuring standard (kWh/kg)

X_{korr} Correction Factors taking into account necessary secondary processes that cannot be eliminated by BNAT.

The best result that is theoretically achievable is 100%. In reality, the energy index can only be approximated, 100% will never be reached. This is highly desirable, as reference parameters do not need to be changed over time. Correction factors may be introduced during the implementation phase if needed.

4.5.6 Energy Efficiency Classes

The main goal of the classification is to divide the possible range into realistic and differentiated classes. If the range of one class is defined too wide, many or all of the facilities may qualify for this class leading to a lack of differentiation between more and less energy efficient machines. It is also desirable to make sure that the majority of the current machines do not qualify for the best class already, so that there is an incentive for innovation (i.e., most of the current refrigerators qualify for the energy label A at the moment).

Due to the definition of the energy index as a division of the minimal theoretical energy consumption it is easy to implement the label using a comprehensive quantitative value.

Therefore the highest energy class should be reserved for BNAT. The next class should reflect the current state-of-art in the field followed by a number of classes containing the less efficient machines.

If letters are used as class names, the classification could look like that shown in **Figure 4.5**.

Energy classes	Type of technology
A	BNAT (Best not available technology)
B	BAT (Best available technology)
C	State of the technology
D	Below state of the technology
E	Old state of the technology
F	Worst available technology

Figure 4.5 General classification of facilities into energy classes

This classification may be used for any industry products. During the introduction phase of the label it should not be possible to reach the highest energy class. This will give an incentive for technologic innovation and an increase in energy efficiency to the manufacturers. The label is not supposed to just mark the current state-of-art in the field.

The definition of the class range is more complex. A first approximation can be done if data from measurements of at least two facilities of different technological levels are available. In **Figure 4.6** the energy classes for PVC profile extrusion are defined.

Energy classes	Type of technology	Measument results	Energy Efficiency Index [%]
A	BNAT (Best not available technology)		> 62
B	BAT (Best available technology)	Argos 93 58.7%	57 - 62
	State of the technology		52 - 57
D	Below state of the technology		47 - 52
E	Old state of the technology	CMT 80 L 42.7%	42 - 47
F	Worst available technology		<41

Figure 4.6 Definition and classification of PVC profile extrusion lines into energy classes

4.5.7 Label Design

The label design chosen is of a hexagonal shape. For easy identification the energy classes are coloured like the EU energy label for the refrigerator. The title of the label was defined as 'Energy Efficiency in Plastics Processing'. Additional important information on the label is: the name and the type of the machine, the type of product, the output as well as the processed material are indicated in a general information section.

Examples for the general information, polymer extrusion:

- Machine name: Argos 93 P
- Machine type: window profile extrusion
- Product types: window profiles, PVC profiles from dryblend
- Guaranteed output: 380 kg/h
- Material: PVC – dry blend

The results of the measurements according to the respective measuring standards (i.e., EUROMAP 60) are indicated in the table with the light blue background (see **Figure 4.7**).

The resulting EEI and the classification are shown in the previous example and were based on the data obtained during the measurements. The label for the Argos 93 P therefore looks like the one in **Figure 4.7**.

4.5.8 Market Introduction and Communication

Further efforts are needed to successfully introduce the label on a European level. For this two options can be considered. In the first option the energy efficiency label will be further developed together with the EUROMAP. The first attempt has already been started in the development of the EUROMAP 30 for energy measurements for extruders. As this is just in the beginning phase of developing a energy measurement standard, it is too early to develop a standard for energy efficiency labelling and needs further awareness raising at the manufacturers and the agents of the EUROMAP.

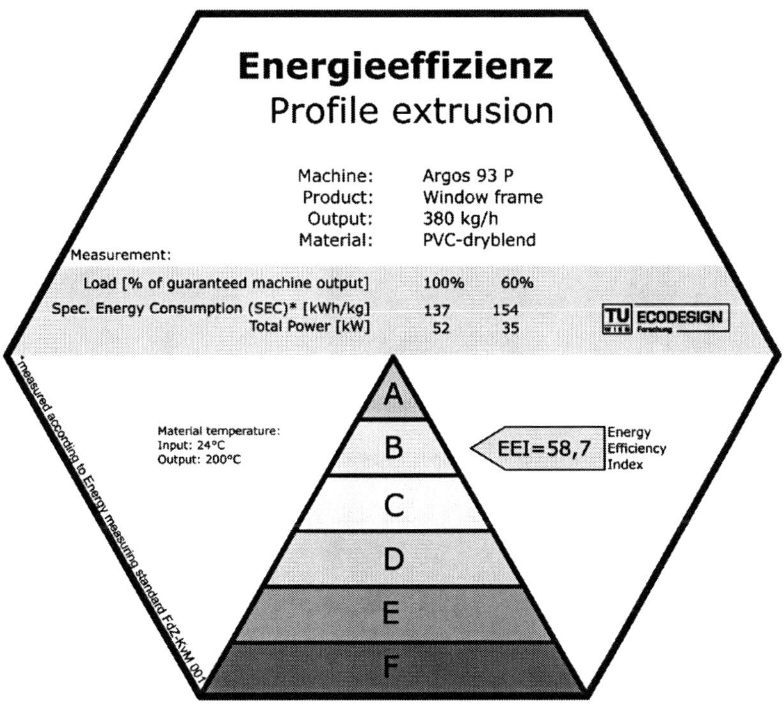

Figure 4.7 Example of an energy efficiency label for the Argos 93 P, Cincinnati Extrusion

4.6 Product Improvement and Ecodesign

With a label in place the challenge is to improve one's products to continuously achieve higher rankings and to market these energy efficient machines with the labelling 'proving' the efficiency performance. To ensure continuous improvement and integrated product development, the product development procedure should incorporate Ecodesign. A suitable method is the Ecodesign Toolbox (*http://www.ecodesign.at/toolbox/*) developed by the research team from the Vienna University of Technology.

The application of the Ecodesign Toolbox helps to systematically gather product characteristics and to identify the life cycle phase with the greatest environmental impact. By using different software tools (*http://www.ecodesign.at/pilot/*) product and process improvement measures can be derived and a Green Product Concept formed.

The requirements of the environment and the stakeholders such as customers and regulatory specifications should be included in the product improvement and lead to sustainable products which will then be communicated to the market. The Ecodesign Toolbox includes the seven process steps shown in **Figure 4.8**.

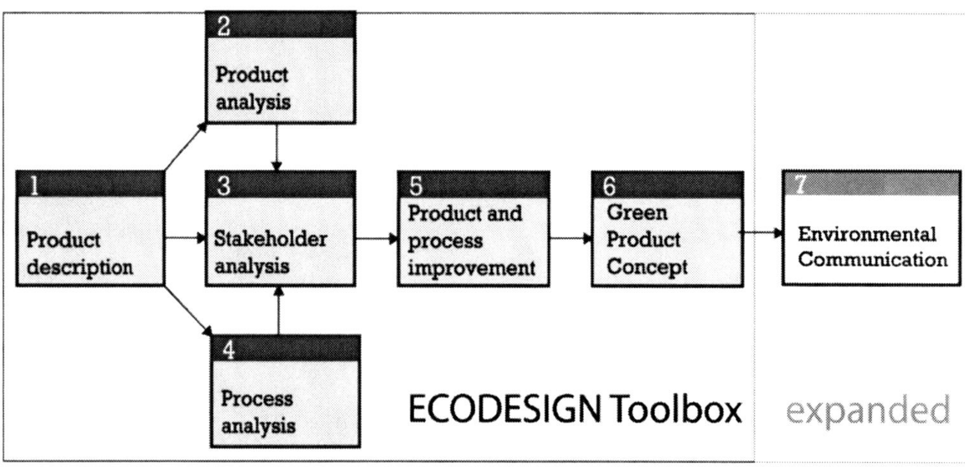

Figure 4.8 Ecodesign toolbox [10]

The product analysis shows that the most energy intensive components are the drive and the heating (**Table 4.1**). These components are therefore most relevant for energy efficiency improvement.

Product improvement step (5) together with the results from the analysis steps (2 and 3) gives a list of improvement measures for reducing the energy demand of the extrusion machine.

Table 4.1 Improvement measures for plastic extruders

Measure	Realisation time	Energy savings potential
Drive		
Use of efficient alternating current drives	Short term	High
Optimise auxiliary drive	Short term	Low
Optimal drive size	Medium term	Medium
Fast running systems	Long term	Medium
Machines without gearbox and so on	Long term	Medium
Heating		
Insulation of the plastification cyclinder	Short term	High
Insulation of the intersections and so on	Short term	High

4.7 Summary

The newly provided energy efficiency label has the potential to raise awareness about energy efficiency in the plastic manufacturing industry. It contributes to a comparison of different machines based on their energy performance supporting decision making of customers. Having a good understanding about the actual energy consumption can also stimulate innovation at the manufacturers looking for new ideas to reduce the energy demand of their machines or the entire system, including upstream and downstream processes.

Certainly the basis for all this is a good standard for measuring the energy consumption. For injection moulding machines the new version of the EUROMAP was an important step to have such a basis for comparison.

The authors hope for wide acceptance of the energy efficiency label not only to reduce the energy demand but also to ensure competitive advantage.

References

1. R. Pamminger, W. Wimmer, M. Huber and M. Bauer, *Die Ökointelligente Spritzgießmaschine*, ECODESIGN, Spritzgießen 2008, Innovation und Produktivität, VDI Verlag GmbH, 2008.

2. EUROMAP 60, *Injection Moulding Machines, Determination of Specific Machine Related Energy Consumption*, Version 1.0, Technical Commission of EUROMAP, 1995, 4.

3. J. Vettkötter, Kunststoff-Berater, *Spritzgießmaschine mit Ökointelligenz*, S. Nachgefragt, Giesel Verlag GmbH, Hannover, Germany, 2008, 5, 46.

4. S. Wiel and J.E. McMahon, *Energy-Efficiency Labels and Standards: A Guidebook for Appliances, Equipment, and Lighting*, 2nd Edition, Lawrence Berkeley National Laboratory, Berkley, CA, USA, 2005. *http://www.escholarship.org/uc/item/01d3r8jg*

5. Directive 2003/66/EC of 3rd July 2003 amending Directive 94/2/EC Implementing Council Directive 92/75/EEC with Regard to Energy Labelling of Household Electric Refrigerators, Freezers and Their Combinations. *http://eur-lex.europa.eu/LexUriServ/LexUriServ.do?uri=OJ:L:2003:170:0010:0014:EN:PDF*

6. GfK Panelmarket, *Verkaufte Kühlgeräte nach Effizienzklassen in den Jahren 2000 bis 2007*, Nuremberg, Germany, 2008.

7. A. Vierling, *Methoden der Wirtschaftlichen Trocknung*, Plastiale Vienna, 2008.

8. EUROMAP 60, [illegible] Technical Commission of EUROMAP, 07/2009.

9. R. Pamminger, W. Wimmer, M. Huber and M. Bauer in the 5th *International Symposium on Environmentally Conscious Design and Inverse Manufacturing*, ECODESIGN Toolbox for the Development of Green Product Concepts, Tokyo, ECODESIGN, 2007.

10. W. Wimmer, R. Pamminger and E. Sehnal, *Development of Criteria for the Communication of the Energy Efficiency of Plastic Processing Machines*, Programmlinie: Fabrik der Zukunft, Endbericht, 2010.

11. EUROMAP 30, *Data Related to Energy Consumption*, Draft, Version 1.0, Technical Commission of EUROMAP.

5 Comparative Analysis of the Carbon Footprint of Wood and Plastic Lumber Railway Sleepers in Brazil and Germany

Ana C. Nioac de Salles, Luiz F. Loureiro Legey, Luiz Pinguelli Rosa, Elen B. Acordi Vasques Pacheco and Jörg Woidasky

5.1 Introduction

The global population growth in the last decades has caused an ever increasing consumption of materials and natural resources, which consequently generated more solid waste, impacting negatively on the environment and requiring a sustainable waste management system. Besides building waste, these discarded materials are mainly composed by plastic, glass, paper, metal and food waste. Plastics are used in various sectors, and in the case of semi- and non-durable consumer goods (e.g., cleaning product containers) have a brief period of use, which accelerates their discard. The constant growth in the volume of plastic waste requires that all countries find a solution to the final destination of these wastes. In general, however, public administration officials have to deal with many issues, including environmental, social and economic matters and do not prioritise policies regarding the final destination of wastes, especially when there is a lack of physical space for the construction or expansion of landfill as well.

The plastic waste management in Germany came into force in 1991 with the Packaging Ordinance, which established that manufacturers and distributors are responsible for their packaging, and to meet specific recycling quotas for their packaging material. Despite that, in many developing countries, for example Brazil, only a small part of the population has access to selective collection programs and a large amount of the solid waste is still disposed of in open dumps.

Another global environmental issue is the increase of deforestation. In the past, deforestation has changed the global geography, when the now developed countries destroyed almost all their forest cover. Unfortunately, it seems that some developing countries are taking the same route and have been cutting their forests at an accelerating pace, as is the case of the Amazon rainforest.

Therefore, one alternative is the reuse of waste materials, which at the same time avoids the cutting of trees. This leads to the need of intensifying the mechanical recycling of post-consumer plastics. Products made of such synthetic materials can be very resistant and durable, as in the case of railway sleepers, where wood is replaced as the raw material. To assess the whole life cycle of every sleeper and identify the positive and negative aspects of each raw material, an environmental analysis is mandatory.

This is the purpose of the study reported in this paper. A comparative analysis of the environmental performance of wood and plastic lumber railway sleepers in Brazil and in Germany, especially their carbon footprint, was carried out in the context of a co-operation between the Fraunhofer Institute ICT and the Energy Planning Program of the Federal University of Rio de Janeiro. The analysis of the whole life cycle of the sleepers was based on the life cycle assessment (LCA) technique and on the International Organization for Standardization ISO 14040. In addition, the exchange of experiences between two countries - a developed and a developing one - made it possible to observe some indicators of their waste management systems and railway sleeper markets.

5.2 Waste Management System

This section presents a comparison between Brazil and Germany, in terms of: (1) composition of households' waste; (2) portion of population served by waste collection programs; (3) quantity of plastic recycled; and (4) annual per person consumption of plastic.

5.2.1 Brazil

The survey conducted by Ciclosoft Research on behalf of the Brazilian Business Commitment for Recycling about the composition of the collected household waste (valuable as recycled materials) found that in 2007, plastics represented 22% (in mass) of total waste. This amount is slightly smaller than paper and cardboard (39%) wastes, as can be seen in **Figure 5.1**. In addition the survey found that only 14% of Brazilians (about 26 million inhabitants) have access to selective collection programs and just 405 out of 5,564 municipalities (or 7%) have a selective collection program [1].

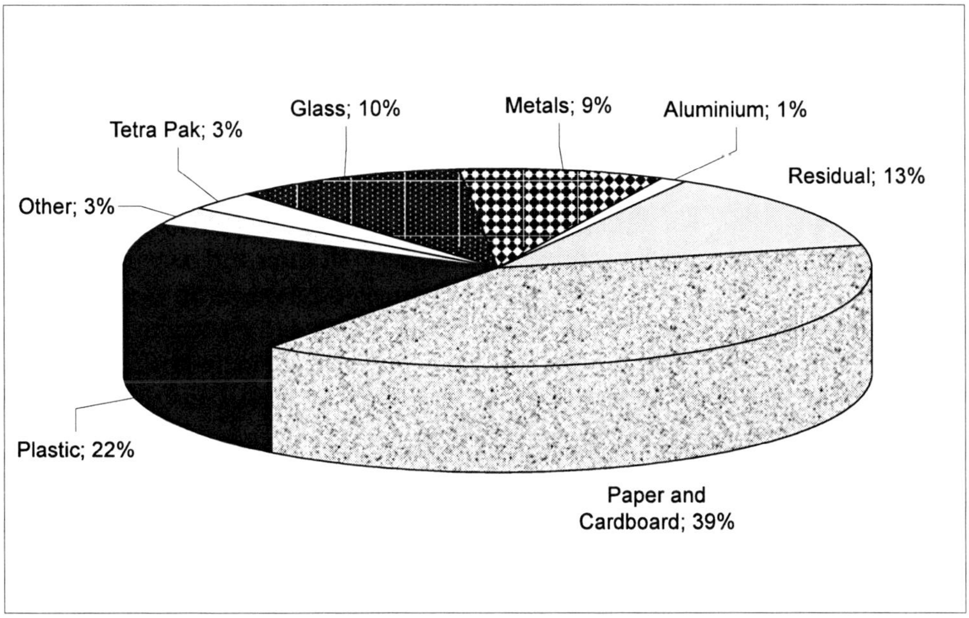

Figure 5.1 Composition of households' waste collected in 2007. Data taken from *Brazilian Business Commitment for Recycling (Cempre), Ciclosoft, Research 2008 - Analysing Selective Collection*, 2008 [1]

In 2007, from the total waste gathered by collection programs - of which 78% were disposed of in open dumps or landfill sites - only 21.2% of plastics were recycled [2].

According to PlasticsEurope Market Research Group [3], the per person consumption of plastic in Latin America was of 21 kg in 2005, as compared to a world average of 30 kg - and which is expected to increase to 32 kg by 2015.

5.2.2 Germany

Manufacturers and distributors of (plastic) packaging materials must ensure a reduction of plastic waste to the environment, by direct take-back schemes at the point of sale, or alternatively by participating in a dual system.

Dual systems first appeared in Germany in 1990, as the Der Grüne Punkt - Duales System Deutschland GmbH (DSD; The Green Dot). In order to liberate industrial firms and retailers from their individual take-back and recovery obligations under the German Packaging Ordinance, DSD was established. Based on this legislation, a second disposal system, complementary to the public-sector waste disposal service, providing collection of used sales packages such as yoghurt tubs, ketchup bottles, beverage cartons and obtaining raw materials from them for the closed-cycle economy. This complementary service includes the collection and recovery of sales packages and facility waste disposal service (thermal and mechanical recycling) and deposit clearing. The tasks involved in collecting, sorting and recovering these used packages are financed by participation fees paid by the manufacturers who put them into circulation. The fees are graduated considering the material and weight of the packages concerned (lighter packages mean lower fees).

As of today, some of these so-called 'dual systems' are operational in the entire country, allowing trade and industries to transfer their take-back responsibility to a private organisation. This action provides financial incentives for improvements and for development of the plastic recycling technology [4].

In 2007, the household waste generated in Germany was 41,750 thousand tons, of which 12% were from packaging (including plastic). However, residuals still represent the largest quantity of waste, 33%, as is illustrated in **Figure 5.2**. Almost all of the population (about 82 million) has access to the collection system [5].

According to PlasticsEurope [3], the final destinations of the plastic waste collected in Germany in 2007, were the following: mechanical recycling (35%); energy recovery (60%); and landfill/storage (5%). In 2005, the consumption of plastic in Western Europe per person was 99 kg and it is expected that this amount will reach 136 kg by 2015.

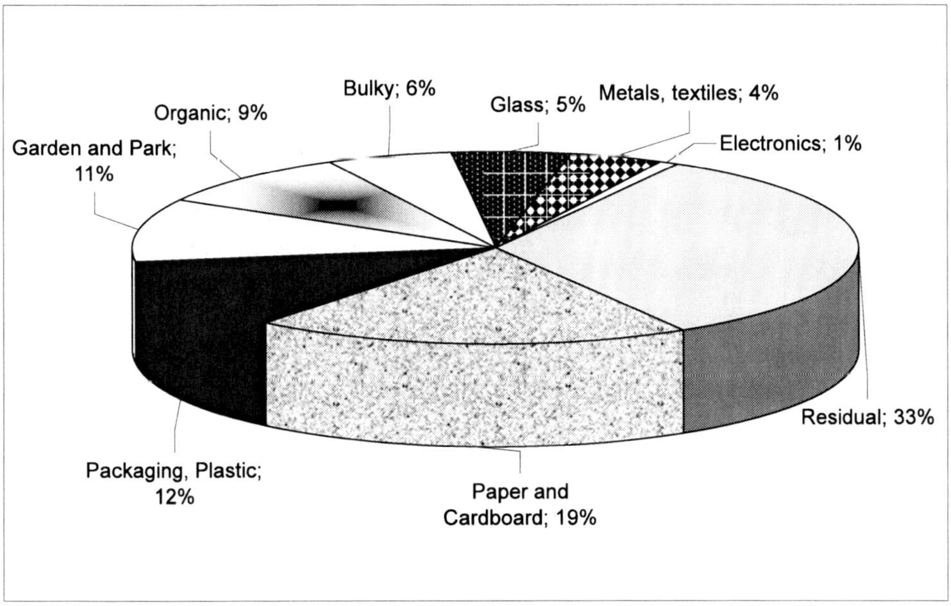

Figure 5.2 Composition of households' waste collected in 2007. Data taken from Destatis, *Umwelt – Abfallbilanz 2007*, Statistisches Bundesamt, Wiesbaden, Germany, 2009 [5]

5.3 Railway Sleepers Market

Railway sleepers are used in railway tracks to support and fix the rails, as well as to transfer loads and vibrations from rails to the ballast (the gravel that fills the space between sleepers) and to maintain rails at the correct gauge (the distance between the two rails forming a railroad track). Some of the indicators of the Brazilian and German railway sleepers markets are presented next.

According to Wisewood a Brazilian producer of plastic lumber railway sleepers, some of the main advantages of these sleepers when compared to wood sleepers [6], are their expected lifetime and their performance in terms of some mechanical properties. In Brazil, wood sleepers need to be replaced in average every 15-20 years, while the plastic lumber sleepers are expected to last 40-50 years. In addition, they present the following properties that make them an interesting alternative for railway operators: (1) 100% recycled; (2) compatibility with wood sleepers currently used in railway networks; (3) resistance to high loads per axle and high levels of tension;

(4) high resistance to water and biological agents; and (5) resistant to many climatic variations (especially near the sea, as is the case of a great part of the Brazilian railway tracks.)

5.3.1 Brazil

Currently, the Brazilian railway network (32,000 kilometres) has about 60 million sleepers, of which approximately 90% are made of wood, ANTT [7]. Since the beginning of railway track construction, by the end of the 19th century, hard woods from the Amazon rainforest (e.g., Florida holly and *Tabebuia spp*) have been cut for railway sleeper production. As **Table 5.1** shows the lifetime of these wood sleepers is about 20 years [8].

The total annual replacement of sleepers is approximately 4 million units. In other words, 7.4% of the sleepers are changed each year [9].

Table 5.1 Lifetime and price of the railway sleepers in Brazil

	Item evaluated			
	Wood	Plastic	Concrete	Steel
Lifetime (years) - (1)	15-20	40-50	40-50	40
Price (€) - (2)	28-70	59-140	140-175	112-140
Equivalent price per year (€)	1.9-3.5	1.5-2.8	3.5	2.8-3.5

Data taken from *Personal Communication with Development, Production*, Wisewood Commercial and Administration Division, São Paulo, Brazil, 2008 [6]; MRS Logística, Personal Communication with Operation and Commercial Division, Minas Gerais, Brazil, 2008 [8]; and N.M.J. Assad, *Dormentes de Plástico Reciclado*, 3rd Edition, Ed., G. Unicop, Revista e Ampliada, Brazil, 2007 [9].

The current difficulty for monitoring and controlling the legality of the natural wood cut from the Amazon forest - which is supplied mostly by small sawmills - allows for the existence of a profitable black market. According to Faria [10], the hardwood sleepers from the Amazon rainforest, which are illegally produced, cost in the south-east about 15 to 20 US dollars (USD), while the same product from registered companies sells for 40 or even 100 USD (depending on the type of wood and transportation costs). This situation makes it still more difficult to introduce alternative raw materials for railway sleepers. In fact, according to the Institute for Technological Research of São Paulo, the average price of rough sawn native timber in December 2006 was approximately 680 USD/m^3 [11], which implies that the cost of a sleeper with 0.1142 m^3, made of lumber sold legally should be around 75 USD.

5.3.2 Germany

The railway network in Germany totals more than 60,000 kilometres and is the largest of Europe, whose total is around 200,000 kilometres long. There are about 100 million sleepers laying in the German railway tracks, of which approximately 70% are made of concrete, 15% of wood (beech is the most common) and 15% of steel [12].

As wood is not that hard as used in Brazil, in Germany all wood railway sleepers are impregnated with a preservative against biological attack, which here will be considered to be creosote, a substance manufactured by distillation of coal tar. The quantity applied is considered to be 16.2 kg of creosote per sleeper [13]. On the other hand, concrete and steel use a large quantity of energy to be produced [14].

The lifetime of an impregnated wood railway sleeper is about 20 to 30 years [15], while concrete and steel sleepers have a lifetime varying between 30 and 50 years [12], see **Table 5.2**).

According to Thyssenkrupp [17], the annual numbers of sleepers replaced per year is about 11 million (or 11% of the sleepers are changed each year), of which 3 million (or 27%) are wood sleepers and 8 million (or 73%) concrete sleepers.

Table 5.2 Lifetime and price of railway sleepers in Germany				
	Item evaluated			
	Impregnated	Plastic	Concrete	Steel
Lifetime (years) - (1)	20-30	50	30-50	30-50
Price (€) - (2)	49	56-108	31	88
Equivalent price per year (€)	1.6-2.5	1.1-2.2	0.6-1.0	1.8-2.9
Quotation on January 7th 2010: 1.432 USD/Euro Data taken from T. Scheibe, *CFT-Sitzung zum Thema: Eisenbahngleisschwellen aus recyceltem Kunststoff*, Deutsche Bahn AG - DB Mobility, Networks and Logistics, Germany, 2008 [16]				

5.4 Scope Definition and Life Cycle Inventory (LCI)

In what follows, the scope of the study reported in this paper is presented, detailing its functional unit, intended audience, product systems, system boundaries and data collection. For both wood and plastic sleepers, the LCI accounted for all materials and energy, both inputs and outputs across the whole life cycle.

5.4.1 Functional Unit

The functional unit is one kilometre of railway track, installed with 1,852 sleepers of both wood or recycled plastic. The service lifetimes assumed for wood sleepers were 20 years for Brazil and 30 years for Germany, and for plastic sleepers, 50 years for both countries.

5.4.2 Intended Audience

The intended audience of this study consists of railway companies, waste managers (both in public service and private companies), regulators, policymakers, consumers and plastic products designers.

5.4.3 Product Systems and System Boundaries

The product systems and the system boundaries were based on the Brazilian and the German case studies, which include material and energy flow (inputs) and emissions to air, soil and water (outputs).

The unit processes and systems boundaries for railway sleepers made of wood are summarised in **Figures 5.3** and **5.4**, and for plastic lumber sleepers in **Figures 5.5** and **5.6**:

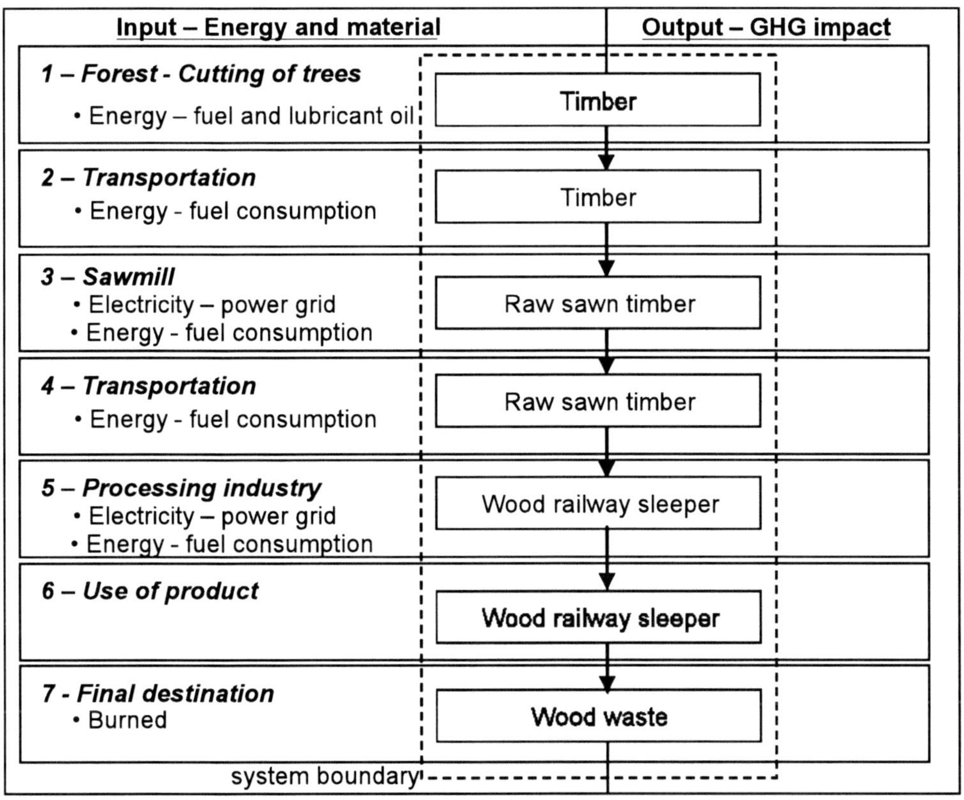

Figure 5.3 Unit processes and system boundary – impregnated wood sleeper in Brazil GHG = Greenhouse gases

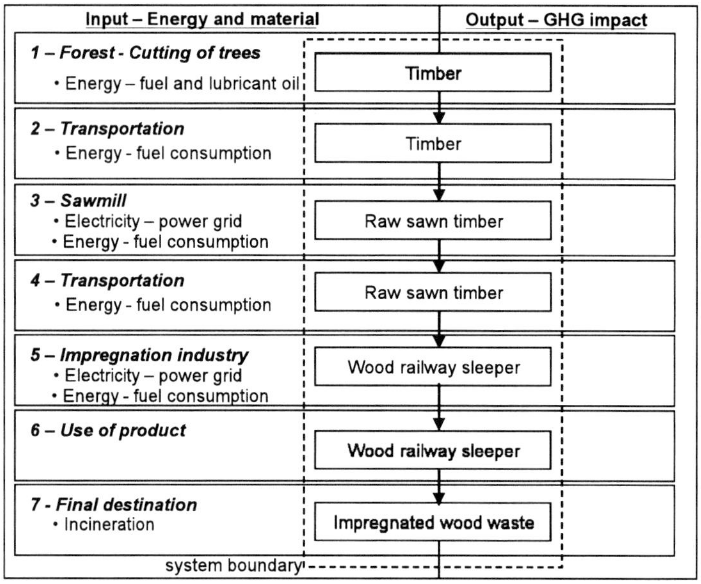

Figure 5.4 Unit processes and system boundary – impregnated wood sleeper in Germany

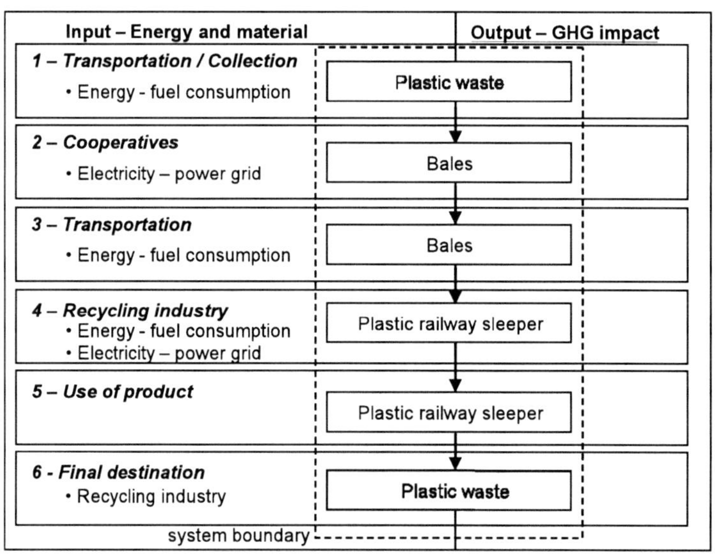

Figure 5.5 Unit processes and system boundary – plastic lumber sleeper in Brazil

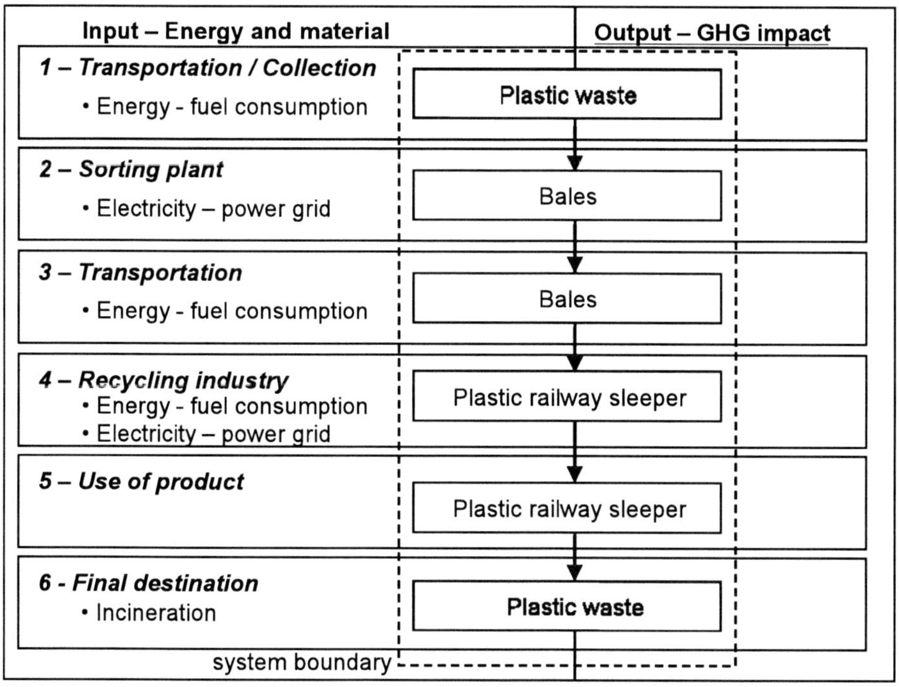

Figure 5.6 Unit processes and system boundary – plastic lumber sleeper in Germany

In the LCI for both German cases, the waste material is assumed to be burned in incinerators. In Brazil, the wood is assumed to be burned in pottery furnaces or used as domestic firewood to produce steam, while the plastic waste is considered to be returned to the recycling industry.

Since the objective of the study is to compare the same product made of either wood or recycled plastic, with basically the same properties, the system boundaries do not include emissions from fuel consumption in the transportation of sleepers to the railway tracks and to their final destination. The reason for this assumption is that this transportation would result in the same emissions for both raw materials. The same holds true for the track bed construction because the structure behind this phase is considered to be almost the same for both raw materials. However, the equipment installed and the logistics on the use phase, which should impact the greenhouse gas (GHG) emissions each time the sleeper is replaced, were also not considered due to the difficulty found in this study to collect this information from the railway companies.

In addition, the energy and the materials used and the waste produced on building sawmills, sorting plants and recycling/processing industries (infrastructure) were not taken into account. Finally, for the wood sleepers, the emissions originated as a consequence of the land use changes brought by the cutting of trees, agriculture or reforestation areas were not taken into account in the system boundary.

5.4.4 Data Collection

The product systems were modeled using information provided by Brazilian and German companies, as well as from the literature and in the GaBi 4 software databases (i.e., plastic packaging in municipal waste incinerator, diesel, power grid mix and thermal energy LCI).

5.5 Results

Results are presented in terms of the functional unit and contribution of each stage of the LCI. A scenario analysis was implemented as well to verify the impact of some parameters on the GHG emissions.

5.5.1 Brazil

Regarding the functional unit defined for this study, to each kilometre of Brazilian railway track a total emission of 1,324 t CO_2 with wood sleepers installed and 170 t CO_2 with plastic lumber sleepers can be allocated. Hence, replacing all the wood by recycled plastic sleepers, the emission of 1,154 t CO_2 could be avoided.

As for wood sleepers, the major GHG emissions are due to the final disposal by burning of wood (81% of total GHG emissions), and to a minor part on the production in a sawmill (14%). For recycled plastic sleepers, the main impacts come from the recycling industry (98%). It is interesting to note that only 0.5% of the GHG emissions comes from the fuel consumption by trucks that collect and transport the raw material to cooperatives and then to recycling plants.

5.5.2 Germany

To each kilometre of German railway track a total emission of 957 kg CO_2 with wood sleepers installed and 839 kg CO_2 with recycled plastic sleepers can be attributed.

Thus, replacing all wood by recycled plastic sleepers, the emission of 118 kg CO_2 could be avoided.

The stages of the lifespan of the wood sleepers that contribute the most to their total GHG emissions are burning of the wood (55%) and the sawmill (37%). For recycled plastic sleepers, the main impacts come from the burning of the plastic (73%) and the recycling industry (20%).

5.5.3 Scenario Analysis

In the **Table 5.3** is presented the GHG emissions for six different final destination scenarios for both wood and recycled plastic sleepers, considering the whole life cycle of each type of sleeper.

Table 5.3 Scenario analysis: GHG emissions

Scenario	GHG Emission (kg CO_2 emissions)			
	Brazil		Germany	
		Recycled	Impregnate	Recycled
1. Considering only the burning of the wood	289	96	321	125
2. Considering the burning of both wood and the plastics for energy recovery (electricity)	122	72	61	-138
3. Considering the burning of both the wood and plastic for energy recovery (electricity), wood as carbon-neutral	-58	72	-119	-138
4. Considering only the burning of the wood and lifetime	724	96	535	125
5. Considering the burning of both wood and the plastic and the lifetime	724	383	535	469
6. Considering the burning of both the wood and the plastic and the lifetime, wood as carbon neutral	256	383	242	469

A scenario analysis was conducted to examine the impact of the lifetime in the GHG emissions over the life cycle of the sleepers, and the sensitivity of different assumptions (e.g., whether to consider wood as a carbon-neutral raw material and the burning of the waste materials for energy recovery).

5.5.4 Brazilian Case

Based on the LCI, this study suggests that the reference scenario for the Brazilian case should be Scenario 4, which considers only the burning of wood and assumes that the plastic waste should return to the recycling industry, to be recycled as long as possible and when not possible anymore, it should be disposed of in landfills. An expected lifetime extension of 30 years, with installation of plastic lumber sleeper, is considered in this scenario as well.

Therefore, considering the Scenario 4 assumptions and the annual replacement of 3.6 million wood sleepers in the Brazilian railroad system, the emissions avoided by replacing plastic lumber for wood in the production of railway sleepers, could be around 2.3 thousand metric tons of CO_2 emissions per year. As is shown in **Table 5.4**, these avoided emissions represent the emissions of almost 31 million Africans.

Table 5.4 Annual emissions per person and the number of inhabitants represented by the emissions avoided by using plastics lumber sleepers instead of wood sleepers

	Africa	Brazil	World	Organisation for economic co-operation and development	USA
Emissions per person (tons of carbon/year)	0.02	0.5	1.14	1.75	5.5
Number of inhabitants represented by the emissions avoided (thousand)	30,755	410	540	352	216

Reproduced with permission from C. Nobre, *The Tictac of the Earth Climate*, National Institute for Space Research (INPE), Brazil, 2008 [18]; Brazilian Institute of Geography and Statistics (IBGE), *Contagem da população 2007*, Brazil, 2007 [19]; and M. Ross and S. Mark, *Stationarity of Global Per Capita Carbon Dioxide Emissions: Implications for Global Warming Scenarios*, 2006 [20].

Another way to visualise the meaning of the avoided emissions through the use of recycled plastic sleepers is by observing that one hectare of the Amazon rainforest can absorb 5 t CO_2 per year [21]. Thus, the 2.3 thousand metric tons of CO_2 emissions that are avoided is equivalent to the annual absorption of a forest with an area four times the size of the city of Rio de Janeiro (1,182 km^2, according to IBGE [19]).

The replacement of wood by plastic sleepers could also avoid an annual cutting of 1.2 million trees, since each native tree produces three railway sleepers. This represents a forest of 30 km^2, or the size of 340 Maracanã Stadiums in Rio de Janeiro (soccer arena with an area of 88 thousand m^2).

As shown in **Figure** 5.7, if the lifetimes of both sleepers were 20 years, the recycled plastic sleeper emissions would be 193 kg CO_2 emissions lower than those of the wood sleeper. This is due to the fact that there are no GHG emissions during the use (service time) of both sleepers. But when the expected lifetime of the recycled plastic sleeper is considered, by the end of 50 years the avoided emissions could reach 628 kg CO_2 emissions, since in that period of time 2.5 wood sleepers are needed instead of just one plastic sleeper. In other words, for the same service unit, only one plastic instead of two and a half wooden sleepers is necessary. Thus, the variation of the GHG emissions stems from the one and a half wooden sleepers that have to be provided for the same service time.

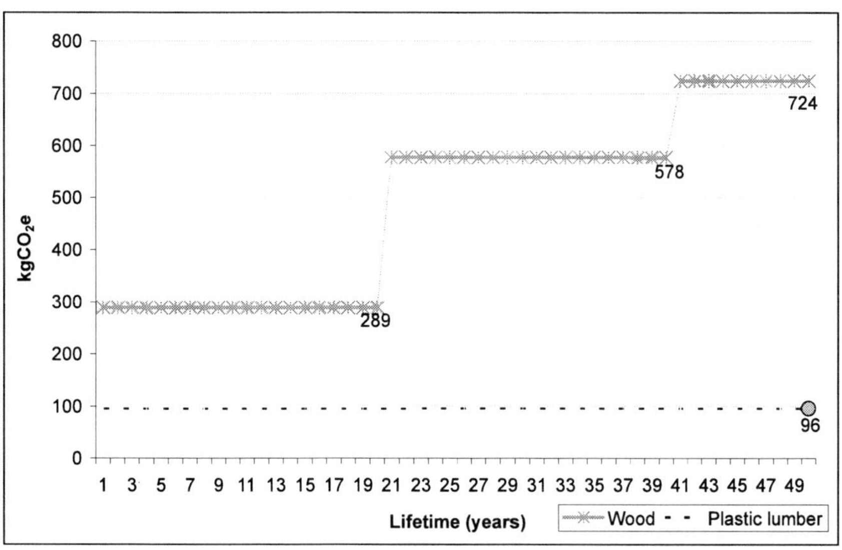

Figure 5.7 GHG emissions during the sleepers' lifetime - Brazil

5.5.5 German Case

Based on the LCI for the German sleepers, this study indicated that Scenario 6 would be the reference scenario, since both materials are burned at the end of their service time for energy recovery. As in the case of Brazil, a lifetime extension of 30 years was assumed, together with a carbon neutral assumption for wood, since this material is extracted from managed forests. In this scenario, if the carbon footprint is considered, if the wood sleepers are replaced by plastic railway sleepers, this would result in an increased production of 227 kg CO_2 emissions. On the other hand, when considering a carbon emission from the wood (Scenario 5), this result changes, and an emission of 66 kg CO_2 emissions could be avoided. Furthermore, since 60% of the plastic waste in Germany has been destined to energy recovery, Scenario 3 could also be taken as the reference scenario, assuming that this waste material is used to replace coal, nuclear and natural gas resources to produce electricity for power grid. In this scenario, the GHG emissions of both sleepers are quite similar.

Considering an annual replacement of 3 million wooden sleepers in the German railway tracks and the fact that two sleepers can be produced from each tree, the replacement of all wood by plastic sleepers would avoid the annual cutting of 1.5 million trees. This represents a forest area of 37.5 km^2, or the size of 586 Allianz Arena - Bayern München Stadiums (a soccer arena with 64 thousand m^2) or 424 Maracanãs to compare with the Brazilian example.

The impact of the lifetime on the GHG emissions besides the carbon neutral issue is illustrated in **Figure 5.8**. If the lifetime of both sleepers were 30 years, the wood sleeper emissions would be 148 kg CO_2 emissions lower than those of the recycled plastic sleeper, since the impregnated wood sleepers emit non-GHG creosote-derived substances while in use. However, when the expected lifetime of 50 years is taken into account, the plastic lumber sleeper could avoid the emission of 66 kg CO_2 emissions, since 1.7 wood sleepers are needed for one plastic sleeper in the same period. In other words, for the same service unit one plastic instead of 1.7 wood sleepers is necessary. Thus, the GHG-delta stems from the 0.7 wood sleepers that have to be provided additionally for the same service time. Finally, it is worth mentioning that the wood carbon neutral assumption reduces the total of the wood sleeper emissions (535 kg CO_2 emissions) by 293 kg CO_2 emissions.

Figure 5.8 GHG emissions during the sleepers' lifetime – Germany

5.6 Discussions and Conclusions

The comparative analysis of the carbon footprint of wood and plastic railway sleepers in Brazil and in Germany brought into discussion not only their potential of GHG savings, but also some other issues related to forest and waste management.

From the inventory it was possible to observe the contribution of each unit process to the total GHG emissions. For wood sleepers in Brazil, sawmill (29% of GHG emissions) and burning of wood (65% of GHG emissions) are the largest emitters. In Germany the results are similar, but the contribution of sawmill (37% of GHG emissions) is greater than in Brazil and the burning of wood (55% of GHG emissions total) is smaller. For plastic lumber sleepers in Brazil, the recycling industry is responsible for 96% of GHG emissions. But, in Germany the results differ upon the consideration of incineration plants as final destination of plastic waste, and the burning accounts for 73% of GHG emissions.

The scenario analysis allowed for the assessment of the impact of some assumptions on GHG emissions. The definition of the reference scenario varies according to the time and place where the product is being produced, installed and discarded. Behind the assumption of the wood as a carbon neutral material there is a big discussion, which involves matters such as forest management and biodiversity preservation concepts, besides the historical responsibility of GHG emissions of each country.

The carbon neutral assumption for wood in Germany means that the emissions caused by the deforestation of native trees, which is accounted as a carbon stock, would be offset by the new forests, planted for commercial purposes. However, this represents actually a flow of carbon in the sense that although current forests absorb carbon from the atmosphere during their growth, when they are cut the carbon returns to the atmosphere, creating a new cycle (closed), which does not compensate for the original forests. The carbon emissions caused by the deforestation of the original forest would only be offset by the planting and conservation of a new forest with native species, re-establishing the original stock, notwithstanding the loss of biodiversity and other aspects that should be considered as well when sustainable system is being discussed. Of course, if commercial forests were planted in areas where a forest cover has never existed, the wood extracted from them could be considered as a carbon neutral material.

The original forests that covered 93% of the German territory 8,000 years ago (about 33 million hectares) no longer exist. The current German mixed forests (377,000 hectares) have been planted especially after the Industrial Revolution and their management has been established for many years. In the case of the Amazon rainforest significant deforestation is recent, though. It began during the 1970s and has accelerated in the past two decades. For this reason, it has been included in the global climate change discussion.

The issue of wood carbon neutrality in Germany involves political negotiation between countries concerning different methodologies to account for historical, current and forecast emissions. Thus, LCA studies should take this fact into consideration. Since it is not straightforward to define whether wood is a carbon neutral material, even when cut from a managed forest - especially because biodiversity and the risk of animal extinction issues are involved - the present study used the technique of scenario analysis to consider different assumptions.

In addition, it should be pointed out that if all the equipment and logistical efforts necessary to the installation of each sleepers were taken into account, the GHG emission of the wood sleepers would be even higher, due to its expected lifetime. Also, there are emissions such as tar oil vapours from the wood sleepers used in Germany, which are affecting the environment, but do not contribute to the GHG and therefore are not taken into account in this study. Creosote consists of polycyclic aromatic hydrocarbons and some of these substances - especially benzo[a]pyrene - have been classified as carcinogenic, thus presenting environmental impact associated with air pollution (photochemical ozone creation) and human toxicity [15].

In terms of waste management in Brazil, because recycling is a source of revenue for the low income population, it basically happens spontaneously. Nevertheless, the replacement of wood by plastic lumber sleepers could add value to materials that are disposed of in landfills and open dumps. In addition it would reduce the volume of waste discarded - thus expanding the length of time of operation of landfills - and contributes to the generation of work and income. The recycling could be even greater if the waste selection and the number of collection facilities were expanded. For recycling to become increasingly attractive economically, the Brazilian Government needs to devise tax incentives and other mechanisms to stimulate investments in waste selection and collection systems, similar to what happens, for example, in Germany. These incentives should include the largest possible number of Brazilian municipalities as well as the recycling industry as a whole. In a country with significant social problems, such as extreme poverty and one of the most heterogeneous income distributions, this would be certainly a welcomed practice. Nevertheless, to conceive ways for improving the selective waste collection in a country with continental proportions — in which 200 million inhabitants cope with huge regional differences, not only social and economic, but cultural and ethnic as well - represents a challenge facing the emergence of a sustainable recycling business.

Thus, from the GHG emissions analysis, plastic lumber sleepers in Brazil are recommended, whereas their use in Germany can be only recommended under specific conditions, such as the reuse of plastic waste as raw material for railway sleepers, which can delay its incineration for at least some decades. In both Germany and Brazil, the availability of plastic waste should increase with the growing of discarded materials in the coming years, which could meet, at least in part, the annual demand for the production of plastic lumber railway sleepers.

Of course, the existence of an alternative to the wooden railway sleepers cannot avoid the cutting of trees, but it may help bring to discussion related environmental issues and promote a reflection on the part of the intended audience concerning future actions. Along those lines, the LCI study presented here intends to provide some information to help a better comprehension of the totality of environmental parameters connected to railway sleepers.

Finally, it is important to mention that the comparison between two countries, one developed and the other developing, may bring a better understanding of how each of them deal with the environmental issues presented here. This may promote the exchange of experiences and help each country to deal with these matters.

Acknowledgments

The authors wish to thank the Brazilian National Oil Agency, Deutscher Akademischer Austausch Dienst and the Federal University of Rio de Janeiro for their financial support. In addition, they would like to express their gratitude to the German Ministry for Education and Research and the funding agency PTJ, and to the Companies: Wisewood, NGR Recycling, PAV Recyclate, Nehlsen, Jakob Becker and Imprägnierwerk Wülknitz.

References

1. Brazilian Business Commitment for Recycling (Cempre), *Ciclosoft, Research 2008 - Analysing Selective Collection*, 2008.

2. Plastivida, *Research of Market of Recycling*, Brazil, 2008.

3. PlasticsEurope, *The Compelling Facts About Plastics 2007 - An Analysis of Plastics Production, Demand and Recovery for 2007 in Europe*, 2008.

4. J.M. Woidasky, *Plastic Recycling in Germany*, Organized by ARDEF and SINPLAST, Brazil, 1999.

5. Destatis, *Umwelt – Abfallbilanz 2007*, Statistisches Bundesamt, Wiesbaden, Germany, 2009.

6. Personal Communication with Development, Production, Commercial and Administration Division, Wisewood, São Paulo, Brazil, 2008.

7. National Agency for Land Transport (ANTT), *Railway Concessions*, Brazil, 2008.

8. Personal Communication with Operation and Commercial Division, MRS Logística, Minas Gerais, Brazil, 2008.

9. N.M.J. Assad, *Dormentes de Plástico Reciclado*, 3rd Edition, Ed., G. Unicop, Revista e Ampliada, Brazil, 2007.

10. A.S. Faria, *Utilização de Materiais Reciclados na Fabricação de Dormentes Ferroviários*, Military Engineering Institute, Brazil, 2006.

11. IPT, *Índice IPT - Preço Médio de Madeira Serrada Bruta*, Institute for Technological Research of São Paulo, Brazil, 2006.

12. Fib, *Precast Concrete Railway Track Systems*, State-of-the-Art Report, Bulletin 37, Fédération Internationale du Béton, 2006.

13. Wülknitz, Contact made directly with the Imprägnierwerk Wülknitz Company, Wülknitz, Germany, 2008.

14. F. Werner, U. Halupczok, R. Schimmelpfennig, H. Willeke, H. Brieke, M. Levering, M-A. Vuilleumier, S-D. Richtberg, J. Dummer, O. Arlt, P. Nowaczyk and R. Schrägle, *Ökobilanzstudie Langfassung*, Vergleich von Schwellen aus Buchenholz, Eichenholz, Beton und Stahl. Studiengesellschaft Holzschwellenoberbau e.V, 2008.

15. M. Kohler and T. Künniger, *Holz als Roh- und Werkstoff*, 2003, **61**, 117.

16. T. Scheibe, *CFT-Sitzung zum Thema: Eisenbahngleisschwellen aus recyceltem Kunststoffe*, Deutsche Bahn AG - DB Mobility, Networks and Logistics, Germany, 2008.

17. Contact Made Directly with the Thyssenkrupp Company, Germany, 2008.

18. C. Nobre, *The Tictac of the Earth Climate*, National Institute for Space Research (INPE), Brazil, 2008.

19. Brazilian Institute of Geography and Statistics (IBGE), *Contagem da População 2007*, Brazil, 2007.

20. R. McKitrick and M.C. Strazicich, *Stationarity of Global Per Capita Carbon Dioxide Emissions: Implications for Global Warming Scenarios*, 2006.

21. National Institute for Space Research (INPE), *Amazon Rainforest Absorbs Excess of CO_2 from the Atmosphere*, Brazil, 2000.

6 Perfect Sorting Solutions for Packaging Recycling

Peter Mayer

6.1 Post-Consumer polyethylene terephthalate through the Ages

In the past couple of years most of the post-consumer polyethylene terephthalate (PET) bottles ended up being the basis for textile-fibre production. The end uses have changed due to new possibilities in the recycling process. The demand for food-grade recycled polyethylene terephthalate (RPET) grew dramatically when soft drink manufacturers like Coca Cola, Pepsi and Nestlé announced that the use of RPET in their bottle production should increase to 60% by 2018. As a legal basis for the use of RPET in production of new packaging, the material has to fulfil the European Food Safety Authority norms in Europe, the need of FDA approval in the US and the need to apply the technical regulations for post-consumer PET in the Mercosur area. These norms guarantee the harmlessness of the recycled product and require highest quality standards of the RPET. There is a very high demand on the recycling process itself due to the inhomogeneous material entering the process. In some countries in the European Union, or in some states in the US, post-consumer bottles are collected via a deposit system or reverse vending machines and therefore have low contamination. There are countries, where the bottles are collected together with other waste materials or are recovered from a landfill. These bottles are getting back into the recycling loop with a high contamination level. These contaminants need to be separated during the recycling process by means of state of the art sorting technology with highest efficiencies.

Basically there are different types of contaminants which need to be separated:

- Foreign objects such as polyvinyl chloride (PVC) bottles, cans, metal contaminates and so on
- Residue from the bottle manufacturing process (labels, glue)
- Contaminants which have migrated into the plastics (bacteria, or for example

aromatic substances caused by incorrect usage of the bottles to store chemical substances or fuel)

6.2 Bottle Sorting, the First Step in the Recycling Process

Depending on the collection system, the post-consumer bottles will reach the recycling plant in bales or loose bottles. The first step is to remove the metal banding, which holds the bales together. This will be cut off and removed manually. The bottles are liberated in a de-baling process from the baled form by means of a 'bale breaker'. Remaining pieces of the metal banding as well as tin cans that are present with the PET bottles will be removed by a high energy magnetic drum separator. As there is a considerable amount of shear force during the de-baling process, some labels will also be removed.

In the ballistic separator, light weight, large scale, labels and foils will be transported upwards due to the paddle movement. The PET bottles slide down due to the angle of the separator and are transported to the sensor-based sorting steps.

The first sorting machine will inspect the flow of bottles for any coarse non ferrous metal contaminations (See **Figure 6.1** bottle sorting process).

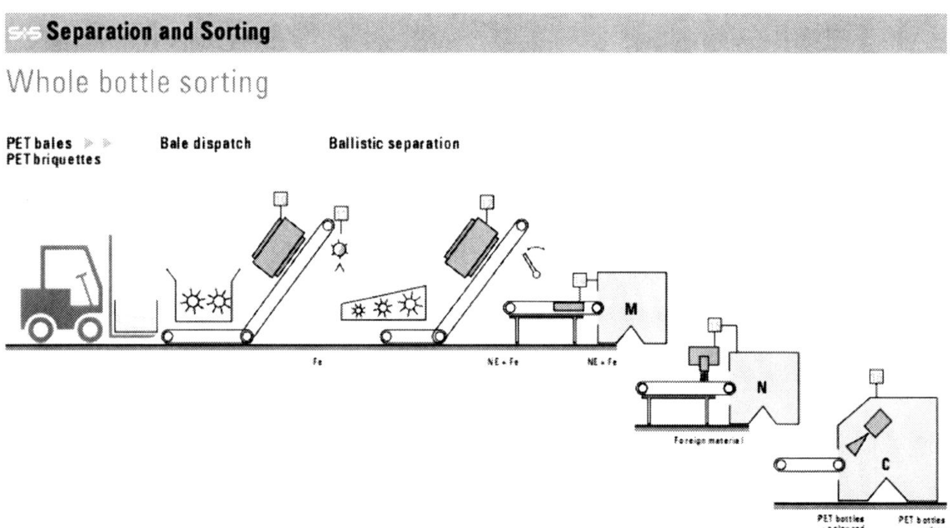

Figure 6.1 Schematic of the bottle sorting process

The bottles are fed onto a high speed conveyor. Beneath the conveyor is a sensor bar, consisting of 200 independently working sensors. The sensors emit a high frequency electromagnetic signal. A receiver coil inside the sensor receives and analyses the emitted signal. If metal contaminants of any kind pass the sensor, due to their conductivity the signal amplitude will be changed, this change will be registered by the receiver coil. The location of the piece is registered by means of a micro-processor. The processor makes a decision based on the precise location of the metal, and energises one of up to 200 pneumatic solenoid valves located at the end of the conveyor. This blast of air removes the material from their original trajectory into a separate discharge bin. This sensor technology is an 'All Metal Sensor', ferrous and non ferrous metals, stainless steel or even metals bonded to a plastic bottle or inside a plastic bottle can be detected and rejected. To avoid a loss of PET material at this stage, the sensor can be adjusted so small metal particles, like remaining pieces of aluminium rings located at the bottle neck, will be ignored in order to avoid the rejection of these bottles into the waste stream. Larger metal contaminants like aluminium cans or composite cans made of PET with an aluminium top however will be rejected in order to avoid the metal from being crushed into many small pieces later in the process.

In the next step, foreign plastics which are different in their chemical composition to PET need to be rejected. Very often these bottles show little or no difference in their optical properties, therefore they can only be identified by using near infrared spectroscopy in non visible optical band. The bottles are inspected by illuminating them with a broadband light unit. The reflection of the light will be evaluated by a special sensor at multiple wavelengths at the same time. Depending on their chemical composition the bottles absorb the light at certain wavelengths; this effect can be used as reliable information to classify the plastic types. Due to the high spectral resolution of 4 nm and 128 different wavelengths evaluated at the same time, the sorting result is more precise than the results of a conventional infrared sensor based system. By rejecting foreign plastic types like PVC or bioplastics like polylactic acid, a contaminant free re-granulate with constant physical and chemical properties can be achieved.

In the last bottle sorting step, depending on requirements the bottles can be sorted by colour. The bottles are fed onto a high speed conveyor and will pass a high speed charged couple device camera system. Intelligent image processing allows to ignore dust and dirt and to reduce false tracking of the bottles due to huge labels to a minimum. Programming of the sorting machine is done by an 'Auto Learn' programme, where an operator trains the machine by simply presenting the bottles to the sensor. This allows previously unidentified colours to be entered into the machines database with ease to the operator.

Bottles coated with titanium dioxide (TiO_2) to prevent ultraviolet radiation will be rejected as well at this stage. For as little as one TiO_2 bottle in a whole bale will cause some apparent discoloring in the new bottles.

Separation and sorting bottle sorters of the 'Varisort' type (S+S Separation and Sorting Technology GmbH) are built in a modular fashion. If space is a restriction, multiple sensors can be combined in one machine and multiple sorting steps can be realised with one machine.

6.3 Quality Improvement and Decontamination during the Flake Washing and Sorting Process

In the next step of the recycling process the pre-sorted bottles will be ground using either a dry granulation process or a wet shredding process, followed by a hot washing to remove labels or remaining glue from the PET flakes. To enhance the removal by having additional friction between the flakes in some plants, friction- or turbo-washers will be used. To remove caps made of polyolefin the ground flakes will be fed to a flotation tank, where lightweight polyolefins float and heavy PET sinks to the bottom of the tank before it is discharged. Remaining metal contaminations which could cause metal inclusions in the pre-forms (see **Figure 6.2**) will cause the pre-form to burst in the blow mould process, and will be rejected in the following step.

Figure 6.2 Bottle pre-form with metal-inclusion

The flake inspection is accomplished by feeding them onto a vibratory feeder that spreads them evenly where they cascade in a thin layer down a glass chute. Beneath this glass chute is a segmented high-frequency detector which can detect metal pieces as small as 0.6 mm. If metal is detected it will be rejected by a blast of air. In this step, mainly metals are rejected which would have caused an enormous material loss before grinding due to the fact that they are quite small and are bonded to the PET (see **Figure 6.3** Flake Sorting process). For the former treatment of the flakes there are basically two different ways: By means of special treatment processes like the American United Research Recovery Corporation (URRC) process or the process invented by German machine manufacturer Krones, the washed flakes will be directly upgraded to food grade flakes. To achieve this high quality, the flakes will be heated and their upper surface (which could be contaminated by substances which did migrate into the plastic bottle), will be cleaned off by friction using a chemical dissolver. Another option to decontaminate the flakes later on in the process is by re-pelletising the flakes which will be explained more in detail later on. In the next step, the flakes will be sorted by colour. Discolored flakes as well as flakes with remaining glue will be rejected, as well as polyolefin parts that were not removed during the flotation process.

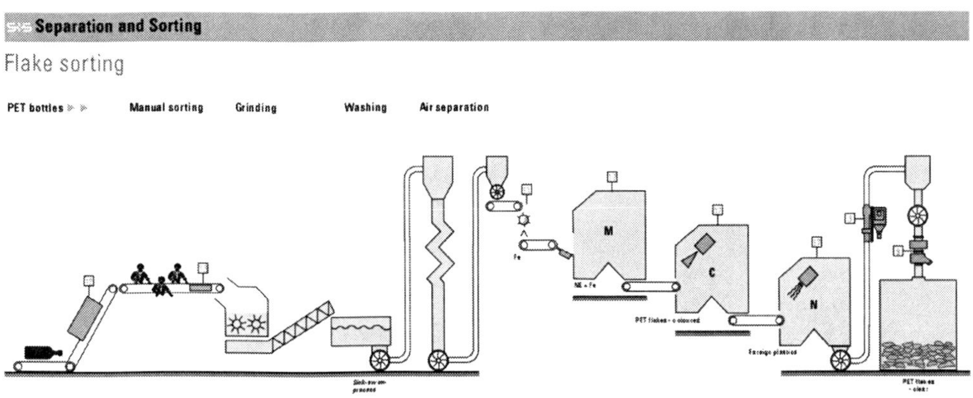

Figure 6.3 Schematic of the flakes-sorting process

By using true colour cameras which can determine between 256 million colours, it is possible to separate for example light green flakes from clear flakes, whereas the light blue flakes still remain with the clear flakes which helps to brighten the colour of the clear flakes. If there is some thermal treatment in the previous processing some plastics like PVC or polyamide (which is very often used as a barrier layer in multilayer bottles) change colour due to heat influence and therefore can easily be separated by the colour separator. If the decontamination takes places during the re-pelletising process, there is no discoloring of that material and therefore, no chance to separate it by means of the colour separator. In this case, these contaminates as well as contaminants that do not change colour in a heating process, (for example silicone seals found inside the caps of sports drinks), will be separated in the final step, in the flake sorting process with a polymer type separator Purifier *N*. Similar to bottle sorting, the detection of the foreign plastics is based on a multi-spectral-near infrared sensor which can distinguish between different plastic types based on their light absorption at different wavelengths. Up to 180 million flakes will be analysed by this sensor per hour, and up to 480 high precision ejection valves will reject the contaminants with up to 250,000 air impulses per second. Again similar to the bottle sorting machines, the flake sorter is designed in a modular construction, and therefore more sorting steps can be done in one machine if necessary (see **Figure 6.4**).

Figure 6.4 Flake Purifier CMN – combined colour, metal and plastic type sorter for flakes

The decontaminated flakes can be directly processed on specially designed extruders to produce preforms. In a process without a direct decontamination in the flake stage, the flakes will be decontaminated by means of special reactors and extruders produced by the Austrian manufactures Erema or Starlinger. During pelletising, the flakes will be cleaned and their physical properties like intrinsic viscosity will be increased. This granulate will be used for preform-production. Other than using this for bottle production, more and more food grade PET material will also be used for sheet production or the production of trays.

6.4 Bottle to Bottle Recycling - The Ecological Alternative

From an ecological point of view, a closed bottle to bottle loop is more preferable than adding additives to the PET bottles which enable the bottle to biodegrade, as the energy which has been used for bottle production, can be reused in most of the part. PET bottle to bottle recycling (with the exception of glass recycling), is one of the few recycling processes where the recycled products have the same value as the original, and therefore no 'down cycling' is in the process.

Bibliography/Further Reading

1. *http://www.pet-recycling.com*

2. *http://www.erema.at*

3. *http://www.starlinger.com/en/recycling/*

4. *http://www.krones.com/*

5. *http://www.urrc.net*

6. *http://www.polyester-technology.com*

7. *http://www.pet-recyclingv.net*

7 UK Household Plastic Packaging Collection Survey 2009

Stuart Foster

7.1 UK Household Plastics Packaging Recycling Survey Background

The information collected for the 2009 survey is based on 2008 data and forms the basis of this document, which is the *15th Local Authority Plastics Collection Report*. The purpose of this report is to inform stakeholders including local authorities, waste management companies and reprocessors on current plastic packaging collection methods and practices. The report also details collection rates and the future plans of local authorities in respect of plastics recycling.

In March 2009, Recoup distributed an online Household Plastics Packaging Collection Survey to all UK local authorities to ascertain the levels of plastics packaging collection in 2008. As in the past the reaction to the survey was extremely positive with 380 responses from Waste Collection Authorities (WCA) and a further 24 responses from Waste Disposal Authorities (WDA). The information submitted and the associated analysis within this report will be of particular relevance to those associated with local and national government together with those in industry who have interests and responsibilities in developing sustainable plastics packaging recycling from the domestic waste stream. Where only partial data has been received or no data submitted Recoup has, as in previous years, completed an estimated data set based on a number of factors including householders serviced by plastic bottle collection schemes, bring sites available and previous year's survey responses, or in some cases by applying reasonable assumptions and average performance data.

7.2 UK Plastic Packaging Consumption Statistics

The key survey data is based on comparing reported household plastic packaging recycling tonnage with consumption estimates. This data needs to be corrected to reflect best estimated packaging trends and ensure that 2008 bottle recycling rates are not overstated. In 2006 a detailed review of plastics packaging consumption was completed. This indicated that 548,000 tonnes of plastic bottles were consumed in 2005. It was expected that up to 23,000 tonnes was consumed outside of the home, leaving 525,000 tonnes entering the domestic waste and recycling systems.

Packaging trends are frequently debated with between 0% and 5% per year growth indicated. This report has applied a 2% annual packaging growth rate for plastics. This is inclusive of any packaging reduction projects, light weighting activities, and substitution of other materials. Therefore, the actual expected UK consumption rate for plastic bottles in 2008 was 581,500 tonnes of which an estimated 557,000 tonnes entered the household waste stream.

7.3 Household Plastic Packaging Recycling Rates in 2008

The analysis of the 2008 survey data indicates that a total of 238,768 tonnes of plastic packaging was collected from bring and kerbside schemes for recycling by the waste collection and disposal authorities. Of the 238,768 tonnes of plastic packaging collected, 216,067 tonnes was reported as plastic bottles. The data as shown in **Table 7.1** shows an additional 34,180 tonnes of plastic bottles collected for recycling when compared to the 181,887 tonnes for 2007, or a 19% increase in year on year bottle collections.

Table 7.1 Breakdown of Recycled Tonnage by Scheme and Packaging Type

	Plastic Bottles (tonnes)	Non Bottle Plastic Packaging (tonnes)	Total (tonnes)
Bring	44,181	3,910	48,091
Kerbside	168,996	18,791	187,757
Recycle 'on the go'	2,920	0	2,920
Total	**216,067**	**22,701**	**238,768**

The total UK plastic bottle consumption within the household waste streams for 2008 and used in this report is 557,000 tonnes indicating a recycling rate of 39%. Despite adjustments made in consumption data, this still demonstrates an increase compared with the previous year. An additional 2,920 tonnes of bottles were collected from recycling 'on the go' services.

This year local authorities were asked to differentiate their collection of plastic bottles between 'bring', 'kerbside' and recycle 'on the go' schemes, with the latter becoming more prevalent since the DEFRA consultation document in 2007. This survey has also sought to split the levels of plastics collected between plastic bottles and other household plastics packaging, such as pots, tubs and trays. **Figure 7.1** shows the historic levels of plastic collected and also provides some inferred future recovery levels based on the data received.

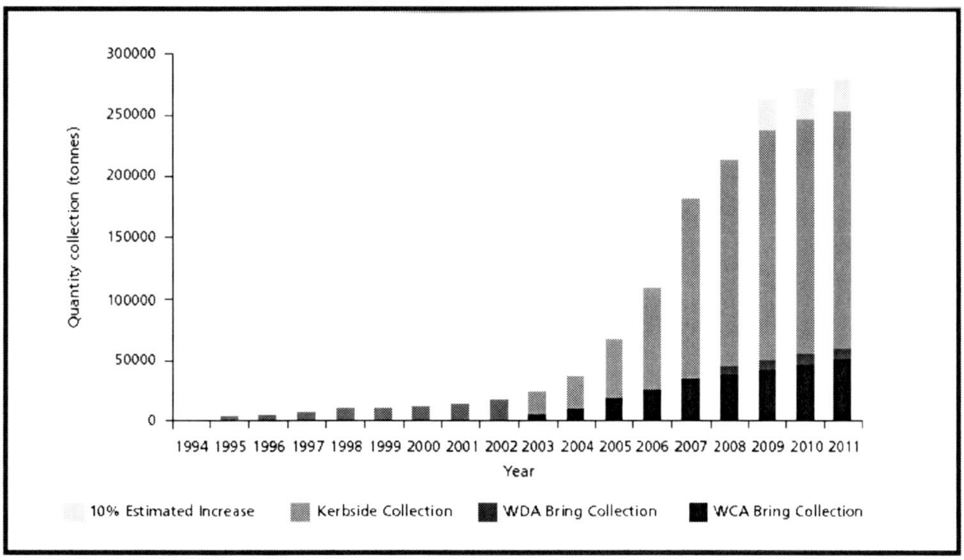

Figure 7.1 Growth in Plastic Bottle Collections

7.4 Plastic Bottle Collection Infrastructure Summary

Table 7.2 shows the breakdown of plastic bottle collections by country and infrastructure. The two main collection schemes offered to householders by the local authorities are kerbside collection schemes and bring schemes.

Table 7.2 Breakdown of Plastic Bottles Collections by Country and Infrastructure

			Bring Schemes		Kerbside Schemes	
Country	Households (millions)	Local Authorities (LA)	Number of LA	Number of Sites	Number of LA	Number of Households (millions)
England	21.1	395	226	5192	253	14.8
Northern Ireland	0.7	26	14	121	25	0.6
Scotland	2.3	32	17	1319	25	1.8
Wales	1.2	22	14	114	18	0.9
Total	25.3	475	271	6746	321	18.1

England accounts for 83% of plastic bottles collected through the bring schemes and 82% of bottles collected through the kerbside schemes. The data indicate that there has been growth in the plastic tonnage collected by the local authorities in 2008. Whilst in certain areas there has been significant levels of growth, for Wales, the amount of plastic bottles collected through the bring schemes has dropped when compared against the 2007 data. This is almost certainly attributed to the increase in the amount of material now collected through the kerbside schemes, a factor which is likely to affect all bring schemes in future years.

7.5 Bring Scheme Performance

This year's data identified that the current bring collection tonnage for plastic bottles is now 44,181 tonnes including 5,801 tonnes from household waste recycling centres operated by waste disposal authorities. An additional 2,920 tonnes of bottles was attributed to recycle 'on the go' activities.

The total number of bring sites actively collecting plastic bottles is now reported as 6,746. The 2007 data indicated that approximately 34,500 tonnes of plastic bottles were collected through the bring schemes suggesting a 28% recovery increase for 2008.

7.6 Kerbside Scheme Performance

In 2008, 168,966 tonnes of plastic bottles were collected through kerbside schemes. This shows an increase of 21,561 tonnes on the previous year – representing a 13% increase. This increase has occurred due to a number of local authorities introducing kerbside collection for the first time, the expansion of some schemes and overall performance improvements of existing services. The total number of local authorities now offering plastic bottle recycling through kerbside collections has risen from 304 in 2007 to 321 in 2008. Kerbside schemes in 2008 provided collection services for some 18 million households representing an additional 4 million households on 2007.

7.7 Reported Perceptions of Running Plastic Bottle Collections

A total of 209 local authorities provided a response on how their plastic bottle collection scheme was considered to be working. A number of authorities (160) indicated that their schemes were running smoothly. A further 46 suggested that there were minor problems with the schemes, whilst three stated that there were major problems in running a plastics recycling scheme. The overall response indicated that the infrastructure and support required to launch and sustainably operate a plastics collection scheme is available and understood by most local authorities. As part of the survey, those local authorities not collecting plastic bottles via kerbside were asked to indicate the main reasons which prevented them offering this service. The most common reason given was the cost implications of changing existing kerbside schemes to collect plastic bottles. The second most common reason was the fact that the local authorities have a focus on collecting the heavier recyclables in order to achieve their recycling rates and as a consequence of this, the collection of lighter waste fractions which can be recycled have lower priority.

7.8 Collection of Non Bottle Plastics Packaging for Recycling

The 2009 survey requested additional information on the types of materials collected by local authorities. Of the 380 responding local authorities, 49 indicated that they are now actively collecting other plastics through kerbside schemes in addition to plastic bottles. A further 35 local authorities indicated that they were collecting 'other plastics packaging' through their bring schemes.

The total reported tonnage for 'other plastic packaging', increasingly being referred to as 'mixed plastics', and typically including pots, tubs, trays and plastic film (but not plastic bottles) was 22,701 tonnes. This includes 1,505 tonnes being collected by WDA through household waste recycling centres. Whilst this reported figure is accepted and incorporated into the overall collection data, for 2008 there is known to be some schemes collecting 'mixed plastics' which are then baled with plastic bottles. This would impact on the overall individual tonnages reported for plastic bottles and mixed plastics.

The information provided in the survey suggests an increase of mixed plastics on the 2007 data of 109%, but this is based on a relatively low 2007 figure and the inability to accurately identify, in some cases, what is the level of non bottle plastics packaging being collected. It is reasonable to believe that the 22,701 tonnes figure reported for 2008 is understated. **Table 7.3** provides a more detailed breakdown of the number of collection schemes which now include various non bottle plastic packaging formats. Greater use of material assessment checks are needed to quantify the extent of non bottle plastics in baled bottles.

Table 7.3 Comparison between 2007 and 2008 Data for Non Bottle Plastics Packaging Collections

	Carrier Bags		Packaging Films		Pots, Tubs & Trays		Dense Plastics	
Other plastics collections	2008 Data %	2007 Data %	2008 Data %	2007 Data %	2008 Data %	2007 Data %	2008 Data %	2007 Data %
Bring schemes	6%	7%	5%	3%	10%	5%	4%	2%
Kerbside schemes	9%	6%	6%	2%	19%	7%	2%	1%
Total schemes	15%	13%	11%	5%	29%	12%	6%	3%

7.9 Sale of Material

The collection and recycling of plastics from the domestic waste stream remains primarily focused on plastic bottles. Local authorities tend to target the collection of all bottle formats e.g., high-density polyethylene and polyethylene terephthalate bottles both in clear/natural and in various colours. These are sometimes segregated by material type and colour before baling. In the last two years UK markets for non-segregated baled bottles have developed as new Plastics Reclamation Facilities (PRF) have been commissioned. However a high proportion of non-segregated baled bottles are still exported to the Far East. Demand for this product is strong with suppliers unable to meet commercial demands. Of those schemes which are collecting 'other' plastic packaging in addition to plastic bottles, the over-riding majority indicated that whilst the material was collected in a mixed format, they were intended to be sold to market as a lower grade of bottles. The development of PRF has allowed this approach to expand. No materials reclamation facilities identified at the time of the survey confirmed their ability to segregate bottle fractions from other plastic packaging formats.

7.10 Planned Developments

7.10.1 Bring Schemes

Since 2005, the collection of plastic bottles through bring scheme sites has experienced a year on year growth of approximately 20%. This year's data indicates that 44,181 tonnes of plastic bottles have been collected through the bring schemes, which represents an increase of 28%. What is clearly visible from the information received is that bring schemes are a supplement rather than an alternative to kerbside collections if the key aim is to increase the tonnage of plastic bottles collected. The reported developments of bring scheme sites are nominal during the next three years. Therefore, greater than 40,000 tonnes of bottles can be expected from bring schemes on an annual basis until 2011. The WDA were also asked for their plans until 2011 for the collection of plastics. The data indicates that 289 new sites will be introduced for plastic bottles over this period. The new sites will nominally increase the overall plastic bottle recovery levels as many will be introduced in areas which have a kerbside or bring scheme service through the WCA.

7.10.2 Kerbside Schemes

The recovery of plastic bottles as part of kerbside collection schemes has seen dramatic growth since 2005. This year, kerbside collection has experienced a further growth of 13% to 168,966 tonnes. Kerbside schemes continue to grow as a result of many factors. New collections are being launched ever year, existing schemes are being expanded upon, and more local authorities are now adopting good practices to maximise system efficiency and tonnage recovery. The growth estimation was based on reported plans from local authorities as well as a 10% additional increase to account for population growth and local authorities that did not have plans confirmed at the time of the survey.

The data regarding new kerbside collections suggests that an annual bottle collection rate of approximately 194,000 tonnes can be expected by 2011. This should be considered as indicative only as some local authorities were not able to provide details at the time of the survey. If the yearly growth in number of households receiving kerbside plastic bottle collections can be increased by a further 10% per year from 2009 onwards, the quantity of plastic bottles collected from kerbside in 2011 would increase to over 257,000 tonnes. If a yearly 10% increase in overall kg/household/y performance across all schemes was also realised, the estimated plastic bottle collection would be over 342,000 tonnes by 2011. This information is illustrated in **Figure 7.2**.

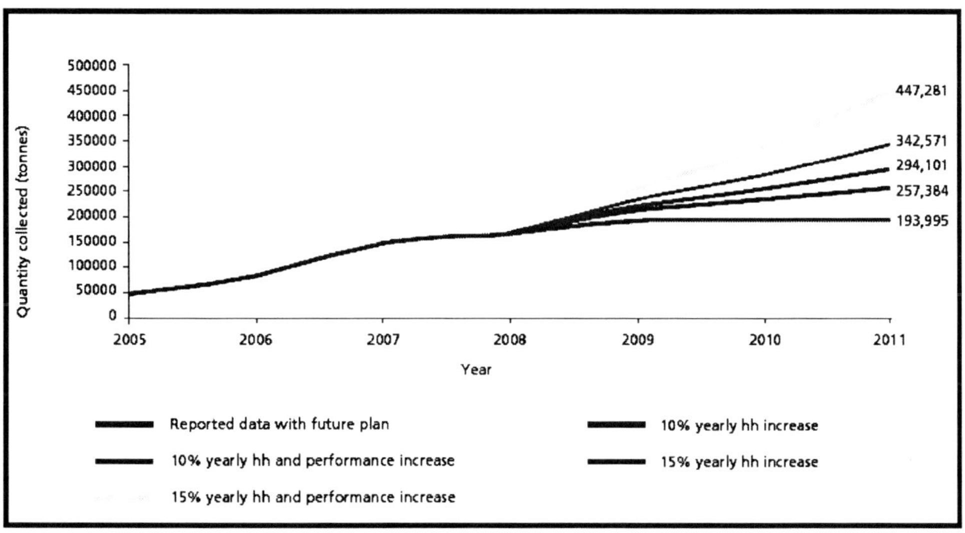

Figure 7.2 Kerbside Plastic Bottle Collection Growth Estimates based on Actual Data and Inferred Infrastructure and Performance Increases

7.11 Development of Non Bottle Plastics Packaging Collections

There is now an increasing trend from local authorities to include 'other' plastics packaging in the kerbside recycling collection scheme. It is more difficult to gauge the development of mixed plastics recycling, and therefore future collection rates are also difficult to predict. While future increased tonnage of 'other' plastics packaging is expected, the scale of this increase will be largely dependent on market values, and the growth in handling and recycling opportunities.

In 2008 there were a number of new reprocessing and sorting facilities which became operational, increasing the capacity to process more plastics packaging commonly found in the household waste bin. As this infrastructure continues to change, the ability for local authorities to collect more material will occur.

Bibliography/Further Reading

The information provided in this section was produced from local authority and waste management data collected and analysed by Recoup. The next edition of the Survey 2010 report will be available in Autumn 2010 from the Recoup website.

1. Recoup (Recycling of Used Plastics Limited). *www.recoup.org*

2. European Plastics Recovery and Recycling Organisation. *http://www.epro-plasticsrecycling.org/*

3. British Plastics Federation Recycling Council. *http://www.bpf.co.uk/about_the_bpf/bpf_recycling_council.aspx*

4. Waste and Resources Action Programme. *www.wrap.org.uk*

8 Vinyl 2010: Experience and Perspectives in Polyvinyl Chloride (PVC) Sustainable Development

Arjen Sevenster

8.1 PVC: Strengths and Concerns

PVC is one of the most widely used plastic materials. It is durable and fulfils challenging requirements in versatile applications. About half of its volume has a service life of more than 35 years - close to 70% of European PVC resin production is used for construction. PVC is also used for medical devices, packaging, shelter against adverse weather conditions, clothing, protection for electrical cables and many other useful applications.

Concerns have been voiced for many years regarding potential impacts of its production, additives and waste management. These have motivated PVC industry to thoroughly investigate these issues and take appropriate measures.

8.2 The Vinyl 2010 Initiative

8.2.1 Vinyl 2010: Foundation, Structure and Organisation

In March 2000, European PVC raw material producers and converters signed an agreement '*Vinyl 2010 – The Voluntary Commitment of the PVC industry*. In doing so, they committed to complete a 10 year plan with quantitative targets and clear deadlines, in order to improve production processes and products, minimise emissions, and boost waste collection and recycling, in brief, enhancing sustainability of PVC throughout its entire life cycle.

Vinyl 2010 is also the legal entity set up to provide the necessary organisational and financial infrastructure and manage this Voluntary Commitment. Vinyl 2010 is an International Non Profit Association with three members:

- European Council of Vinyl Manufacturers (ECVM)
- European Stabilisers Producers Association (ESPA)
- European Plastics Converters (EuPC)

ECVM, ESPA, EuPC, representing the producers of PVC resin, of stabilisers, and the converters of plastic materials respectively. The European Council of Plasticisers and Intermediates (ECPI) participates to the management and financial support of Vinyl 2010 without formally being a member.

A Monitoring Committee was set up early on to verify the progress of the programme, and guarantee good governance, effectiveness and transparency in the implementation of the Voluntary Commitment. It makes recommendations on strategic issues and includes senior representatives from the European Commission (EC), the European Parliament, consumer organisations, trade unions, and industry.

Vinyl 2010 publishes annually a transparent and thorough Progress Report featuring verified and detailed results as well as financial information.

Independent auditors ensure the validity of financial accounts and achievements stated in the Annual Progress Report to ensure they truly represent Vinyl 2010's performance.

8.2.2 Commitments

Within the frame of Vinyl 2010 commitments were made, relating to all phases of the life cycle of PVC products.

8.2.2.1 Manufacturing

- Cut raw materials and energy consumption where warranted economically and ecologically.
- Minimise environmental impact through compliance with ECVM Industry Charters [1].

8.2.2.2 Plasticisers

- Ensure that plasticisers are used safely, in line with the final conclusions of European Union (EU) risk assessments.

- Conduct material lifecycle analyses to identify additional improvements to products and their use.
- Conduct research, adding to the already available scientific data about plasticisers, to help policy-makers in taking informed decisions.

8.2.2.3 Stabilisers

- Phase out cadmium stabilisers in the EU by 2001.
- Compile the necessary data for a full EU risk assessment on lead.
- Develop alternatives to lead-based stabiliser systems, as part of a commitment to replace these by 2015, with clear intermediate targets.
- Produce yearly statistics on stabiliser consumption by converters.

An additional commitment to discontinue use of bisphenol A in the production of PVC resin was implemented in 2001, and formally included in Vinyl 2010 during its 2005 mid-term review.

8.2.2.4 Waste Management

- Actively support integrated waste management approaches, using raw materials as efficiently as possible and developing sustainable end-of-life options.
- Develop and implement by 2010 recycling of 200,000 tonnes per year of PVC post-consumer waste from applications not regulated by EU Directives, in addition to waste already recycled in 2000 in these applications.
- Recycle at least 25% of collectable, available PVC post-consumer waste from pipes, window profiles and roofing membranes by 2003 and increase this to 50% by 2005.

8.3 Activities and Achievements of Vinyl 2010

8.3.1 Manufacturing

Industry's compliance with the Charters for suspension PVC and emulsion PVC was verified by Det Norske Veritas, in 2002 and 2004, respectively.

Compliance with the *suspension PVC Charter* involved 264 measurements in 38 production sites. In 2002, only 3% of these measurements were non-compliant, whilst a further 4% could not be verified.

Compliance of *emulsion PVC* plants was verified in 2004. While almost all plants complied with the criteria for vinyl chloride monomer (VCM) emissions to air and water, full compliance against the limit of one part per million residual VCM in all resin grades proved more challenging. Despite this, approximately 95% of the amount of PVC resin produced during the verification period complied with this criterion.

Emissions follow a clearly downward trend. It is expected that compliance will have increased significantly by 2010, when a further verification of both Charters will take place.

With respect to *raw material and energy consumption* the *Eco-Profiles* for both suspension and emulsion PVC were updated in 2006. Eco-Profiles are life cycle inventories covering all steps, from raw materials production until PVC production. Even though the actual production process for PVC has changed little over the last 20 years, total energy demand has been reduced considerably, as well as the overall environmental impact. The eco-profiles of major polymers are available on the Plastics Europe website (*www.plasticseurope.org*).

8.3.2 Stabilisers

Heavy metals are being phased out wherever possible from PVC stabilisers, a generic group of additives designed to avoid decomposition of PVC during processing and to ensure its long term stability.

Already in 2001, ESPA and EuPC phased out *cadmium stabilisers* in the EU-15.

Vinyl 2010 contributes to *risk assessments* of lead and tin in PVC stabilisers. ESPA is an active partner in the Voluntary Risk Assessment on Lead and Lead Compounds. Conclusions indicate that there is no concern for the general population, but that occupational exposure of workers in stabiliser plants should be further reduced. ESPA members implement a lead in blood reduction programme, aiming at reducing worker's exposure.

Reduction of lead stabilisers exceeded the interim 2005 target of 15%. The commitment to phase out the use of cadmium and (by 2015) of lead in all stabilisers was extended to the EU-25 countries as from 2006. In the period 2000-2008, lead stabilisers (in EU-15) decreased by 66,552 tonnes (-52.3%) (**Table 8.1**).

Table 8.1 Progress of lead replacement

Tonnes of stabiliser systems	2000	2008	Reduction (%)
Formulated lead stabilisers	127,156	60,604	52.3

8.3.3 Plasticisers

For some applications, PVC is rendered flexible by addition of plasticisers. Concern has been expressed regarding the ubiquitous presence of those substances, even though they are barely soluble in water, biodegradable and non bio-accumulative.

In 2005-2006 EU risk assessments were published for three commonly used phthalates: di-isononyl (DINP), diisodecyl (DIDP) and dibutyl phthalate (DBP). These showed that DINP and DIDP pose no risk to human health or the environment in any current application. DBP, however, presents some risk to workers through inhalation and to plants growing close to processing sites. Risks can be managed by air treatment measures and using personal protective equipment.

The risk assessment of di-(2-ethylhexyl) phthalate (DEHP) was published in 2008. It confirmed that adequate risk prevention is already in place for adult consumers. However, there were two concerns identified which required attention: possible risks to patients from the use of DEHP in medical equipment; and possible risks from emissions from converting plants to children living near such sites and from consuming food grown locally, and to aquatic and terrestrial ecosystems.

Regarding medical devices, the EU Scientific Committee for Emerging and Newly Identified Health Risks (SCENIHR) delivered an opinion in March 2008. SCENIHR found that so far, there is no conclusive scientific evidence that DEHP exposure via medical treatments has harmful effects in humans, although further studies should be undertaken.

Despite the hypothetical nature of concerns about DEHP emissions from converting plants, ECPI and EuPC have worked together with the Directorate General Enterprise of the European Commission to identify safe emission levels. Implementation will be imposed as part of the authorisation procedure.

Recommendations about toys were superseded when Directive 2005/84/EC prohibited the use of DEHP in toys and child-care articles from January 2007. The Risk

Assessment Report also identified some concerns relating to medical equipment, investigated further by the appropriate EU Scientific Committee.

The risk assessment for butylbenzyl phthalate was published in 2008.

Risk assessment reports are available on the European Chemicals Bureau websites (*http://ecb.jrc.it* and *http://phthalates.com*).

8.3.4 Waste Management

8.3.4.1 Collection and Recycling for Specific Applications

Collection and recycling schemes were originally set up to deal with waste streams arising from specific applications, mirroring the activity of PVC converting industry. This structure was later adapted where appropriate, to enhance efficiency, by the creation of Recovinyl (**Section 8.3.4.3**).

Schemes for *pipes* now cover EU-15, except Greece and Ireland. *Window profiles* are also collected, except for Greece and Portugal. In these applications supply is limited by long lifetimes, by waste exports, and by re-use of PVC window frames being replaced by better ones. Demand therefore exceeds supply.

Originally, recycling of *roofing membranes* was limited to the German market. Waste was processed in a dedicated cryogenic plant. A new collection scheme was launched in 2003 and extended to the Benelux, Austria and France (2004), and later to Italy, Spain and the UK. The cryogenic plant was closed down - recycling is now carried out by conventional mechanical recycling or by the Vinyloop process (**Section 8.3.4.4**).

Flooring recycling started in Germany, in a dedicated plant which is still in operation. Collection was later extended to Austria, Switzerland, Belgium and France and the UK. A R&D programme focused on new applications for recyclate, produced in the Vinyloop plant. Coating, as well as calendering, has been successfully demonstrated.

Recycling of composites, e.g., a *coated fabric* is still challenging. Waste fabrics have been collected in France since 2002, in Germany since 2004. A specific recycling technology, Texyloop®, is based on the Vinyloop process. A pilot plant was recently completed, as an extension of the Ferrara plant. Conventional mechanical recycling into automotive parts has been taking place since 2004. The technology involves mechanical shredding, partial separation of polyester fibres, and compacting (or agglomeration) of post-consumer waste fabrics.

8.3.4.2 Mixed PVC Recycling

A Vinyl 2010 project ascertained whether mixed PVC waste could be used as filler for *lightweight concrete* products. Preliminary tests carried out in Italy (2004) showed that light concrete with recycled PVC could compare in technical specifications to similar products made with polystyrene or clay, for example:

- Non-structural parts such as floors, roofs and walls, where low weight with high thermal and sound insulating properties are desired
- Refilling road trenches to pass cables or pipes

After a positive feasibility study, the project optimised the blending of light concrete, improving its mechanical, thermal and insulating performances. The Light Concrete project is currently on hold, due to limited availability of suitable waste, but is ready for implementation.

8.3.4.3 Recovinyl

The difficulties encountered in collecting post-consumer waste led to the creation of a new entity, called Recovinyl, which gets its funding and strategic guidance from Vinyl 2010. Its objective is to maintain a steady supply of post-consumer PVC waste for recycling, offering sustainable solutions for collection and recycling of mixed post-consumer PVC waste, essentially from the building and construction sector, by collaborating with established third-party organisations, including municipalities, waste management companies and recyclers.

Recovinyl provides financial incentives to support the collection of PVC waste. Incentive payments encourage recycling by compensating extra costs of collection and recycling, compared to disposal. In cooperation with waste collectors, Recovinyl promotes best practice and shares know-how regarding recovery and recycling solutions. Recovinyl supports waste sorting by developing guidelines and organising training sessions for collection centres, where staff is trained to identify and sort PVC waste.

Recovinyl identifies suitable recycling facilities for PVC waste. To be accredited, these must meet specific requirements and accept that their facilities, processes and operations be audited. Financial assistance to accredited recyclers and waste recovery firms is provided only after proof of receipt and data, such as tonnage, origin and date of shipment are supplied. The waste needs to be accepted by the recycler and to correspond to Recovinyl's selection criteria. Recycling operations are subject to independent auditing commissioned by Recovinyl.

Initially, Recovinyl concentrated its activities in the UK, Belgium and the Netherlands. Recovinyl is now active in fourteen countries – Austria, Belgium, the Czech Republic, Denmark, France, Germany, Hungary, Italy, the Netherlands, Poland, Portugal, Slovakia, Spain and the UK.

8.3.4.4 Mechanical Recycling

Vinyl 2010 sponsored developments in recycling PVC waste that formerly was not amenable to mechanical recycling.

The Vinyloop® process was developed by Solvay. In 2001 the first Vinyloop plant was built in Ferrara (Italy) with a financial contribution from Vinyl 2010, essentially in order to recycle PVC cable waste. Vinyloop® uses solvents to separate PVC from fabrics, metals and other impurities and produces high quality recycled PVC (R-PVC) compounds. Considerable efforts were made to improve the on-stream factor of the plant, reduce production costs, use various types of PVC waste and improve the quality of the R-PVC. The Ferrara plant is now operating at full capacity.

The Texyloop® technology is an add-on module of the Vinyloop process, developed for treating coated fabrics. A pilot with a capacity of 2,000 tonnes/year started up in 2008 in Ferrara.

8.3.4.5 Feedstock Recycling

Attempts by the PVC industry to develop dedicated feedstock recycling processes failed. Vinyl 2010 subsequently decided to consider only technologies proven at least on pilot-plant scale, but so far the only one commercially applied is by Dow at its Schkopau site near Leipzig.

Tests showed that this robust process, operating since 1999 to recycle chlorinated waste, can handle large quantities of PVC waste [2]. Large-scale trials proved the concept. In total, 1027 t of PVC waste were processed and the recovered chlorine used on site for producing VCM. Unfortunately the high costs of logistics and treatment limit its application.

8.3.4.6 Energy Recovery

Vinyl 2010 also considers the fate of PVC ending up in waste incinerators with energy recovery and more specifically the recycling of:

- Hydrochloric acid recovered in some energy recovery plants
- Residues from flue gas neutralisation

Some incineration plants recover hydrochloric acid as well as energy. Tests with PVC waste were carried out in 2001 at an incinerator in Hamburg, where hydrochloric acid is recovered from the raw flue gas, processed into a 30% aqueous solution and sold to chemical industry. PVC waste was added to the normal waste stream in order to reach the maximum design level of HCl in raw flue gas. About 500 tonnes of PVC waste were thus processed over a five week period. The trials were entirely successful. Process parameters and the slag and fly ash composition remained unaffected. The level of dioxins in the gaseous effluent remained far below mandatory limits. No traces of abnormal corrosion were detected during the maintenance shut-down that followed the trial.

The **Halosep® process** was developed by the technology subsidiary Watech of Danish cable manufacturer NKT. Watech is now owned by Stena Metall. The process uses aqueous hydrochloric acid to extract and separate chlorides and heavy metals from neutralisation residues of municipal solid waste incinerators. To demonstrate industrial feasibility a pilot plant was built in 2004 with financial support from Vinyl 2010. Trials were completed during 2006. Treatment of flue gas treatment residues reduces amount and hazardousness of waste, transforming a large part of the waste into a marketable product at competitive cost. At present, Stena is looking for partners to build a commercial-sized demonstration plant.

Although developed without support of Vinyl 2010, it is also worth mentioning the Neutrec® process (Solvay). It reduces the amount of residues from neutralisation with sodium bicarbonate by 80-90%, as most is converted into brine, used as feedstock for sodium carbonate plants. The residues are dissolved at a controlled pH and with certain additives. The insoluble part (heavy metal hydroxides, activated carbon and fly ash) is separated by filtration. The brine passes over activated carbon to adsorb any organic components. Traces of heavy metals are eliminated by passing over ion exchange resins. A pilot industrial plant has operated in Rosignano, Italy since 1998 and a 50 kt/y commercial plant was commissioned in France in 2003.

8.3.4.7 PVC Waste Statistics

The quantities of post-consumer PVC waste collected and recycled with support of Vinyl 2010 are audited externally. **Table 8.2** shows the tonnages recycled.

Table 8.2 Amounts of waste recycled with the support of Vinyl 2010 (tonnes)

2003	14,255
2004	18,077
2005	38,793
2006	82,812
2007	149,463
2008	194,950

However, other key data are more difficult to assemble, for example:

- Total amounts of PVC post-consumer waste generated in the EU, and the proportion which is effectively available and collectable. Pipes that remain buried are considered unavailable, for example. Waste from small and scattered sources cannot be collected efficiently.
- Total amount of PVC industrial waste generated and recycled in the EU.
- Amount of PVC post-consumer waste, recycled without support from Vinyl 2010 (and hence not counted by Vinyl 2010), as well as amount re-used and amount exported outside the EU.

Vinyl 2010 sponsored several studies in order to better measure and assess the data given previously.

Already before 2000, EuPC developed a statistical model to estimate the amount of PVC waste generated in the EU, as well as the amount actually available for collection and the amount collectable for technically and economically feasible recycling. This was crucial to generate realistic figures as a sound basis for establishing future commitments for Vinyl 2010.

The model is based on several key parameters about PVC products and their moves in the waste market. The principal parameters are historical production figures of various PVC products since their introduction into the various European markets, average product life times by application and availability/collectability factors.

Based on the experience gained, the model was later updated, taking into account recent production data, corrected growth rates and longer lifetimes in some applications. The results showed a significant decrease in the quantities of post-consumer PVC waste arising, in line with opinions of experts in the field.

In 2006, similar studies were carried out in Hungary and Poland to estimate the trend of post-consumer PVC waste volumes for the period 2005-2020. These studies forecast a growing trend of PVC waste availability in both countries, mainly due to the building and construction sector, with a substantial increase expected for cables and flooring. In 2007, EuPC carried out a study covering Slovakia.

In 2005 Vinyl 2010 commissioned a study on PVC wastes recycling in the EU-15 to the French consultant AJI-Europe. It showed among others that 92% of post-industrial PVC waste was recycled, either on site, or outside the processing plants.

The German company Consultic has been conducting statistical assessments of PVC waste recycling in Germany for several years. Consultic is now also evaluating plastics waste arising, recycled and recovered on behalf of PlasticsEurope for the entire EU. This provides an alternative way to estimate PVC waste arising and recycling.

8.3.4.8 Partnership with Local Authorities

Since 2001 the PVC industry has collaborated with the Association of Cities and Regions for Recycling and Sustainable Resource Management, Plastics*Europe* and European Plastics Recyclers to improve the recycling of plastic waste collected by local authorities.

Pilot trials confirmed that demolition of old buildings yields only negligible amounts of heterogeneous and heavily contaminated plastic waste. Conversely, trade and small to medium enterprises, as well as packaging from construction sites appeared as significant sources of plastic waste. A 'Best Practice' guide for plastics waste management by local authorities was issued in four languages in 2004 [3].

Assessing the Potential of Plastics Recycling in the Construction and Demolition Activities (APPRICOD) was launched as a follow-up project to research and promote collection and recycling of plastic waste from construction and demolition, under the umbrella of the EU Life Initiative in 2003. Vinyl 2010, its members, and Recovinyl provided technical assistance. The main objective was to assess the potential of plastic recycling in the construction and demolition (C&D) sector, which supplies the largest waste stream in the EU. In terms of weight, plastics are a small component, but they represent significantly more in terms of volume.

In addition to assembling information, about 20 pilot projects were carried out (including in Ancona, Barcelona, Brussels and Porto), covering construction, renovation and demolition.

The results of the project were published in a guide available in six languages through the APPRICOD website [4].

8.3.4.9 Other Partnerships

Since 2004, Vinyl 2010 has been a partnership registered with the Secretariat of the United Nations Commission on Sustainable Development. The aim of this is to contribute to the development of effective industrial partnerships and to the exchange of experiences at global level.

8.4 Lessons Learnt

8.4.1 Manufacturing

The key commitments embodied in ECVM Charters were entered even before Vinyl 2010 was founded (1995). Some emission targets today may now look less demanding, yet continuous monitoring and periodic verification (as required by the Charters) have substantially contributed to this progress. For example, from 2000 to 2007 total annual emissions of VCM to the atmosphere and water, expressed in g/ton of PVC produced, have more than halved, whereas combined emissions of dioxins, which were already very low, have decreased further by more than 70%. Most experts and stakeholders agree that present emission levels from VCM and PVC production are no longer a cause for concern.

8.4.2 Additives

Phasing out of cadmium was the endpoint of a steady reduction, going on since the 1990s and driven by industry. Progressive replacement of lead, on the other hand, is more challenging from both a technical and economic point of view. Still, the intermediate 2005 reduction target was exceeded and the 2010 target was already achieved in 2008, but the efforts and cost involved exceeded expectations. The costs of developing, testing and implementing equipment changes have added to several tens of millions of €. Depending on relative costs of lead and alternatives, increases in variable costs have in some cases been substantial.

Duration, volume and expenditure related to risk assessments have also exceeded expectations, connected with stronger public concern, as well as levels of scrutiny required by authorities. On the positive side, PVC-additives are now among the best prepared to submit to the registration, evaluation and possibly authorisation steps, required by the EU Registration, Evaluation, Authorisation and Restriction of Chemicals (REACH) regulations.

8.4.3 Waste Management

The difficulty and cost of collecting post-consumer PVC waste has proven to be a key challenge, throughout Europe. Market penetration and usage of PVC is rather distinct from country to country and so are local initiatives, interest and willingness for action. Additionally, experience has shown that anticipated arising waste volumes were over-estimated.

Nevertheless, quite some post-consumer PVC still lands into the waste stream and Recovinyl must further raise awareness throughout Europe to avoid disposal of retrievable waste streams. Landfill restrictions could help boost post-consumer PVC recycling.

Another issue is the possible impact of REACH on end-of-life recycling. REACH requires high levels of information exchange between the successive steps in manufacturing chains. Yet, this is unachievable for recyclate sorted from end-of-life waste, due to the huge number of different products involved, the intrinsic variability of waste and long life of some products, for which tracing back the original composition is often impossible. Industry is working together with EU authorities to find pragmatic solutions, ensuring the required level of protection without driving recycling industry out of the EU.

A related issue is the presence of additives in waste which were authorised when these products were made, but are now restricted in new products. A study on the recycling of cadmium containing waste has recently been completed by the Flemish environmental Institute for Technological Research.

Experience confirmed that separate collection represents a significant portion of recycling costs. With exception of electric cables, collected by copper and aluminium scrap recyclers, industry typically pays between 50 and 300 €/t for mono-stream collection, depending on application, volumes, system and location. Such costs come on top of manpower invested to set up, monitor, manage and improve collection schemes. This is the reason why joint collection schemes and partnership with waste management companies and local authorities are actively being developed by Recovinyl.

8.4.4 Recycling Technologies

Progress in conventional mechanical recycling is essentially driven by equipment manufacturers. A typical example is sorting. New applications of recycled waste are being developed. Products including at least some recyclate include co-extruded pipes and window profiles, traffic management objects and mats, sound proofing and non-structural concrete elements, honeycombed water treatment units made from rigid PVC sheets, frames for pouring concrete, etc. Some types of flooring are manufactured from recyclate, and internal sections of window profiles often include recycled PVC.

The novel Vinyloop® technology has reached the required levels of productivity and product quality. The Texyloop® version is now successfully dealing with coated fabrics, recovering both PVC and textile fibres.

Feedstock recycling has been attempted repeatedly, but these trials stumbled over technological and/or economic pitfalls. Incineration with heat recovery is still a realistic route to make the best of the heat and HCl generated from PVC, whether mixed with other materials, or unfit for recycling.

8.5 Future Challenges

With less than one year remaining until target date, the main challenges for Vinyl 2010 are:

- Ensuring that all commitments are fulfilled
- Creating conditions for sustainable continuation of current achievements
- Further improving the sustainability of PVC

The remaining commitments are related essentially to:

- Lead replacement
- Recycling targets

The 2008 *lead replacement* rate already exceeded the 2010 target of 50% reduction, compared to the year 2000. Complete phase-out by 2015 is well on track.

The recycled volumes (**Table 8.2**) show the success of Recovinyl; the 200 kt recycling target is achievable.

The political and regulatory framework is expected to facilitate an optimal use of resources. Restrictions in landfilling will stimulate collection and recycling. However, if not appropriately managed, REACH could be very damaging for recycling post-consumer waste.

Ensuring economic sustainability is a necessity. Vinyl 2010 is entirely financed by contributions from companies operating in the private sector. They compete against other materials that do not bear a similar financial burden. Whilst providing seed money is now necessary and tolerable, this financial burden must not place PVC industry at an unfair competitive disadvantage.

Vinyl 2010 is now well placed to assess the costs of various options and choose eco-efficient ones. The cost per tonne collected is steadily decreasing, but the financial burden still ranges from 5 to 10 millions of € annually. Increasing pressure on natural resources resulting from depletion and economic development will stimulate recycling of waste products. The increasing cost and difficulty of landfilling will bring disposal costs closer to the cost of recycling. These trends are not in doubt. Their timing and impact is more difficult to predict.

8.6 Conclusions

The world has changed a lot since 2000. The concept of *sustainable development* is now universally accepted. The remarkable spread of electronic communication has ensured that many issues are now global, concerns move very quickly across borders, and hence responses must be global as well. Concerns about long-term availability of energetic resources have increased significantly. Climate change is now widely recognised as a major challenge. Health and safety aspects of chemical substances are prompting major regulatory changes. Formerly rural regions have become major manufacturing centres, and citizens are demanding more and better products. Unfortunately, poverty and lack of access to basic necessities, such as water and health care, still affect numerous people worldwide.

Relevant literature [5] has compared 'Command and Control' to other possible approaches leaving some more freedom to industry to achieve similar or more stringent objectives. A promising approach is a Voluntary Commitment (VC) taken by industry. Such VC must be credible and their implementation transparent. The burden of monitoring VC should not become a disincentive, however. The PVC industry has accepted the challenge and launched the innovative voluntary approach 'Vinyl 2010' covering the entire industry chain, from cradle to grave.

In its nine years of existence Vinyl 2010 has gained invaluable experience. It has so far met all commitments without help from outside. Commitment of all stakeholders, sustained efforts by contributors from the entire chain remain as necessary as ever. Current challenges may be somewhat different from those anticipated at the start, yet they remain no less daunting. It is necessary to pay close attention to the economics of collecting and recycling, to adapt to changes in the business, political and regulatory environment, and to innovate in order to develop a viable business model. Considering what has been achieved, the PVC industry can be confident in meeting the challenges of sustainable development, but there is no room for complacency.

1. ECVM Industry Charter for the production of VCM and PVC (suspension process), 1995. Industry Charter for the production of emulsion PVC, 1998 (*www.ecvm.org*).

2. R. Bühl in the *Proceedings of the PVC Conference*, Brighton, UK, 2002, p.375.

3. *Good Practice Guide on Waste Plastics Recycling – A Guide by and for Local and Regional Authorities*, ACRR 2004. The guide is available on the websites *www.ecvm.org* and *www.acrr.org*

4. *Towards Sustainable Plastic C&D Waste Management in Europe*, available in six language versions through the *www.appricod.org* website

5. *ANEC/BEUC position on voluntary environmental agreements*, October 2006

Abbreviations

3D	Three-dimesional
aPP	Atactic polypropylene
APPRICOD	Assessing the Potential of Plastics Recycling in the Construction and Demolition Activities
BAT	Best available technology
BNAT	Best not available technology
C&D	Construction and demolition
DBP	Dibutyl phthalate
DEFRA	Department for Food, Environmnet and Rural Affairs
DEHP	Di-(2-ethylhexyl) phthalate
DIDP	Diisodecyl phthalate
DINP	Diisononyl phthalate
DMA	Dynamic mechanical analysis
DSC	Differential scanning calorimetry
DSD	Duales System Deutschland
EC	European Commission
ECPI	European Council of Plasticisers and Intermediates
ECVM	European Council of Vinyl Manufacturers

EEI	Energy efficiency index
ESPA	European Stabilisers Producers Association
EU	European Union
EuPC	European Plastics Converters
FDA	US Food and Drug Administration
FESEM	Filed-emission scanning electron microscopy
GHG	Greenhouse gases
GWP 100	Global warming potential over 100 years
IBGE	Instituto Brasileiro de Geografia e Estatística
iPP	Isotactic polypropylene
LA	Local Authorities
LCA	Life cycle assessment
LCI	Life cycle inventory
MAH	Maleic anhydride
M_n	Number average molecular weight
M_w	Molecular weight
PA	Polyamide
PET	Polyethylene terephthalate
*p*PBM	*p*-Phenylene-bis-maleamic acid
POx	Peroxide
PP	Polypropylene
PRF	Plastics Reclamation Facilities

PVC	Polyvinyl chloride
R&D	Research and Development
REACH	Registration, Evaluation, Authorisation and Restriction of Chemicals
R-PVC	Recycled polyvinyl chloride
RPET	Recycled polyethylene terephthalate
rpm	revolutions per minute
SA	Succinic anhydride
SCENIHR	Scientific Committee for Emerging and Newly Identified Health Risks
SDoE	Statistical Design of Experiments
SEC	Specific energy consumption
SF	Succinyl fluoresceine
T&D	Transmission and Distribution
T_g	Glass transition temperature
URRC	United Research Recovery Corporation
USD	US Dollars
VC	Voluntary Commitment(s)
VCM	Vinyl chloride monomer
WCA	Waste Collection Authorities
WDA	Waste Disposal Authorities

Contributors

Susana Areso

Grupo de Ingeniería de Polímeros, Instituto de Ciencia y Tecnología de Polímeros. CSIC, C/ Juan de la Cierva 3, 28006 Madrid, Spain

Patrick Bailly

Areva T&D AMT, Boulevard de la Résistance, 71040 Macon, France

John A. Colbert

Smithers Rapra Technology, Shawbury, Shrewsbury, Shropshire, SY4 4NR, UK

Emilia P. Collar

Grupo de Ingeniería de Polímeros, Instituto de Ciencia y Tecnología de Polímeros. CSIC, C/ Juan de la Cierva 3, 28006 Madrid, Spain

Stuart Foster

RECycling Of Used Plastics Ltd., Tower House, Lucy Tower Street, Lincoln, LN1 1XW, UK

Jesús M. García-Martínez

Grupo de Ingeniería de Polímeros, Instituto de Ciencia y Tecnología de Polímeros. CSIC, C/ Juan de la Cierva 3, 28006 Madrid, Spain

Luiz Fernando Loureiro Legey

Centro de Tecnologica, Bloco C, Sala 211 - CEP 21949-972, Cidade Universitária, Ilha do Fundão, Rio de Janeiro, RJ 68565, Brazil

Romain Maladen

Areva T&D AMT, Boulevard de la Résistance, 71040 Macon, France

Peter Mayer

Separation and Sorting Technology GmbH, Regener Strasse 130, Schoenberg, D-94513, Germany

Ana C. Nioac de Salles

Fraunhofer-Institut für Chemische Technologie (ICT), Joseph-von-Fraunhofer-Strasse 7, 76327 Pfinztal, Germany

Elen Beatriz Acordi Vasques Pacheco

Cidade Universitáriam Av. Horácio Macedo, 2.030, Centro de Tecnologica, Prédio do Bloco J, CEP 21941-598, Rio de Janeiro, Brazil

Rainer Pamminger

Vienna University of Technology, Institute for Engineering Design, Getreidemarkt 9, 1060 Vienna, Austria

Daniel Piccoz

Areva T&D AMT, Boulevard de la Résistance, 71040 Macon, France

Christophe Prévé

Areva T&D AMT, Boulevard de la Résistance, 71040 Macon, France

Luiz Pinguelli Rosa

Centro de Tecnologica, Bloco C, Sala 211 - CEP 21949-972, Cidade Universitária, Ilha do Fundão, Rio de Janeiro, RJ 68565, Brazil

Erik Sehnal

Vienna University of Technology, Institute for Engineering Design, Getreidemarkt 9, 1060 Vienna, Austria

Arjen Sevenster

Vinyl 2010, Avenue E. van Nieuwenhuyse 4/4, Brussels, B-1160, Belgium

Jesús Taranco

Grupo de Ingeniería de Polímeros, Instituto de Ciencia y Tecnología de Polímeros. CSIC, C/ Juan de la Cierva 3, 28006 Madrid, Spain

Bernard Valette

Areva T&D AMT, Boulevard de la Résistance, 71040 Macon, France

Wolfgang Wimmer

Vienna University of Technology, Institute for Engineering Design, Getreidemarkt 9, 1060 Vienna, Austria

Jörg Woidasky

Fraunhofer-Institut für Chemische Technologie (ICT), Joseph-von-Fraunhofer-Strasse 7, 76327 Pfinztal, Germany

Index

R

S

T

U

V

W

X

Breinigsville, PA USA
03 February 2011
254687BV00003B/1/P

9 781847 354914